Mathematical Elegance

Mathematical Elegance

An Approachable Guide to Understanding Basic Concepts

Steven Goldberg

This book is printed on acid-free paper that meets the American National Standard for Permanence of Paper for Printed Library Materials.

Library of Congress Catalog Number: 2014012384
ISBN: 978-1-4128-5464-1
Printed in the United States of America

Library of Congress Cataloging-in-Publication Data

Goldberg, Steven, 1941-
Mathematical elegance : an approachable guide to understanding basic concepts / Steven Goldberg.
pages cm
ISBN 978-1-4128-5464-1
1. Mathematics--Popular works. 2. Mathematics--Humor. I. Title.
QA99.G65 2015
510--dc23

2014012384

For Bill and Helaine

Euclid alone has looked on beauty bare.

—Edna St. Vincent Millay

Contents

Acknowledgments

I am grateful for the time, energy, and most helpful suggestions offered by Ibtihaj Arafat, Alan Goldberg, Chris Lawrence, Margarita Levin, Michael Levin, Elizabeth Mayers, Michael Mayers, Mitchell Meltzer, David Singmaster, Faith Scheer, Tuck Washburn, Francis Wilson, and, of course, my wife, Joan Downs Goldberg, without whose inspiration and advice this book could not have come to fruition. And special thanks go to my editor, Laura Parise, and to the marvelous staff at Transaction Publishers whose suggestions made this a better book than it otherwise would have been.

Introduction

I was never very good at math, but I have always loved it for the way it all fits together so nicely. Fortunately, it is a peculiarity of human beings that we can love that, perhaps *only* that, which we do not fully understand. And in this you and I are perhaps more like than unlike the very greatest of mathematicians, men whose thoughts represent some of the highest and most profound we human beings have had.

All of us face a universe in which mathematics has forever unfolded in directions we—and even the greatest mathematicians among us—cannot begin to visualize. Thus, our humility is maintained by the knowledge that, possibly as opposed to the physical universe, the universe of numbers goes on everywhere forever in every possible direction and dimension. These numbers contain unimaginably more mathematical truths than we now know, more even than our universe, even if it has eternity at its disposal, can ever know.

In the mathematical universe, any knowledge of truth we will ever have is insignificant compared to all that is true. Facing such a universe, we realize that the difference between us and the very greatest mathematicians virtually evaporates—sort of. These fellows were a *tad* better than we are at finding mathematical truths. And it is the glory of mathematics that many of the deepest truths can be discovered by looking through the tiny window at the tiny portion of the mathematical universe provided by our little world. It is the mathematicians who look through the window and tell us what they see.

This Is All You Have to Know

There are many different sorts of numbers, most of which are primarily of interest to mathematicians and too arcane to concern us in this book. However, a few types of numbers—most of which are already familiar to you—are the subject of this book. They are the following:

Natural Numbers
Natural numbers are what you first think of when you think of numbers: 1, 2, 3 . . . (". . ." means "goes on forever").

Integers or Whole Numbers
Integers and whole numbers are natural numbers, their negative equivalents (–1, –2, –3 . . .), and 0.

A mathematician named Leopold Kronecker once said that "God made the integers; all else was the work of man." It has always seemed to me that God made the primes and the rest of the numbers are merely a mopping-up operation. But who am I to argue with a guy whose moniker was Kronecker?

Prime Numbers
A prime number is an integer greater than 1 that is only evenly divisible by 1 and itself. Two is a prime number (indeed, it is the only even prime number because every other even number is divisible by 1, itself, and 2). Three, 5, and 7 are prime numbers, but 9 (which is divisible by 1, 9, and 3) is not. Primes are the crème de la crème of numbers because they are the mathematical atoms, the building blocks of all other numbers.

Euclid proved that there is an infinite number of prime numbers (i.e., there is always a larger prime number than the one you thought must be the largest). Some people consider this proof to be the greatest intellectual achievement our species has yet accomplished. Other people do not.

If you can figure out an easy way of determining whether a large number is prime (or an easy formula for generating the primes and only the primes), you will become very, very rich.

Composite Numbers
A composite number is any integer that is not a prime number. Four is a composite number because it is evenly divisible not only

by 1 and itself but also by 2. Fifteen is a composite number because it is evenly divisible by 3 and 5. And the aforementioned 9 is a composite number because it is evenly divisible by 3.

0

Some people consider the discovery of 0 to be the greatest intellectual achievement our species has yet accomplished. Other people do not.

Zero is just as much a number as, say, 3. Three refers to any set of three things, such as three elephants, three ideas, or three trees. Likewise, 0 refers to an "empty set," such as no elephants, no ideas, or no trees.

Unlike the other numbers, 0 causes mathematicians great embarrassment, as we shall see.

Negative Whole Numbers

Negative whole numbers are the mirror images of the natural numbers. If you have a hundred dollars in the bank and write a check for two hundred dollars, you will have negative one hundred dollars (–$100) in the bank. The negative whole numbers, zero, and the natural numbers constitute the integers.

Rational Numbers

Rational numbers can be represented by ratios of two integers, for example, 3/4, 4/3, 3/3 (i.e., 1). All integers and fractions composed of rational numbers are rational, as are all numbers whose decimal equivalent is finite (e.g., 1/4, which is 0.25) or repeating (e.g., 1/3, which is 0.3333333 . . .). All of the types of numbers mentioned above are rational.

Irrational Numbers

Irrational numbers cannot be written in the same forms as rational numbers. The square root of two, for example, cannot be represented by any fraction of rational numbers or any finite or repeating decimal expansion.

Square Root

A square root is a number that is another number multiplied by itself (e.g., 2 is a square root of 4 because 2 times 2 equals 4). The other square root of 2 is –2 because –2 times –2 equals 4.

Real Numbers
Real numbers are the rational numbers and the irrational numbers.

"Imaginary" Numbers
Imaginary numbers are multiples of the square root of −1, which we shall discuss later. For every real number there is an imaginary equivalent. The name of these numbers is unfortunate. They are every bit as legitimate as the real numbers.

Complex Numbers
Complex numbers represent the intersection of real and imaginary numbers, for example, $-6 + 4i$.

Other Numbers
There are many other types of numbers—numbers that are consistent with the previous types but also enable us to describe, feel, and find truth in areas of the mathematical universe we would otherwise know nothing about. These include such types of numbers as hyper-inaccessibles, mahlo cardinals, the first hyper-mahlo, the first weakly compacted cardinal, and, of course, the first ineffable cardinal. Unfortunately, time and space preclude our discussing these numbers in this book. (Yeah, that's it; it's time and space that preclude our discussing these numbers in this book.)

Who Is Number One?

Who is the greatest mathematician? Graduate students in mathematics will argue about this forever, but the nominees certainly include Euclid (Egypt, b. ca. 325 BC); Archimedes (Greece, b. ca. 287 BC); Rene Descartes (France, b. 1596); Pierre de Fermat (France, b. 1601); Blaise Pascal (France, b. 1623); Isaac Newton (England, b. 1642); Gottfried Wilhelm Leibniz (Germany, b. 1646); Leonhard Euler (Switzerland, b. 1707); Karl Friedrich Gauss (Germany, b. 1777); Evariste Galois (France, b. 1811); Bernard Riemann (Germany, b. 1826); Jules Henry Poincare (France, b. 1854); and David Hilbert (Prussia, b. 1862).

This list excludes logicians such as Georg Cantor (Russia, b. 1845) and Kurt Gödel (Austria, b. 1906). Mathematicians and logicians are forever arguing about whether the former is a branch of the latter or vice versa. If logic is considered mathematics, then these two are certainly contenders.

1

The Lay of the Land

Well, I Did Not Want to Know That Anyway

Many of the greatest discoveries of the twentieth century are discoveries of the inherent limitations that will forever withhold knowledge from even our smartest descendants.

Gödel's Theorem: For millennia mathematicians dreamt of discovering an all-encompassing mathematical system in which it was possible to prove every mathematical truth. Half a century ago, however, Kurt Gödel proved that every nontrivial logical and mathematical system will possess truths not provable in that system. (Trust me, you do not want to read here *how* he proved this.)

We *can* demonstrate the truths using a more inclusive system, but we cannot prove that a larger system will have truths without going to an even more robust system. (Ad infinitum.)

The dream is dead.

Heisenberg's Uncertainty Principle: There is a limit beyond which full knowledge is impossible. In other words, there are things we can never, *even in principle*, know. For example, we can never learn both the position and velocity of a subatomic particle. We can learn *either* with as much precision as we wish, but the more precisely we know one, the less precisely we can know the other.

Here's why. When we measure something, we use particles to see the thing we measure and to make the measurement. For example, we use photons, particles of light, to see and measure the tabletop we wish to measure. The fact that we use these particles makes no practical difference when we are viewing and measuring tabletops, elephants, or even bacteria; the effect of the tiny photons on the thing we measure is essentially nonexistent and has no more effect than a ping-pong ball thrown at a mountain range. However, when we attempt to measure particles on the tiny scale of the particles we use to measure them, the effect of the latter on the former is tremendous and sets limits on what we can *ever* know about the measured particles.

Unpredictability: Even when a system is determinative, it is often the case that complexity and sensitive dependence on initial conditions renders the system forever unknowable in practice. This is why we will never be any good at predicting weather conditions more than a couple of weeks ahead.

Unsolvable in Practice (Probably): There are a large number of problems that are probably inherently unsolvable in a finite amount of time. Some of these are problems that you would think would be easy pickings for the mathematician armed with a computer. The traveling salesman problem is a good example.

Say you are a traveling salesman who must visit a certain number of cities, and you want to take the shortest route possible. You might think that there is no practical application of mathematics more easily accomplished than this. And, in fact, the problem is easy to solve by trial and error if the number of cities is relatively small.

For any specific number of cities, the number of routes is the number of cities "factorial"; take my word for this. The meaning of "factorial— which is indicated by an exclamation point—is most easily seen by example: 5! means $5 \times 4 \times 3 \times 2 \times 1$, or 120. Thus, there are 120 different routes for visiting five cities one time each.

If you must visit ten cities, the number of routes your trial-and-error methods must compare is over three and a half million. To be sure, you could obviously immediately disregard inefficient routes such as New York to Los Angeles to Boston to Seattle . . . , but there would remain a daunting number of possible routes that would require actual comparison. With the help of a computer you could do this in a reasonable amount of time, but this is for only ten cities.

Perhaps you must visit thirty thousand cities. A mathematician cannot give you the shortest route that reaches all of the cities; he cannot even tell you with certainty whether there is any way other than trial and error for finding the shortest route. There *are* shortcuts that guarantee a route not more than about 20 percent longer than the shortest route, but no known method other than trial and error that guarantees the shortest route.

Now, you may well say, "Who has to visit thirty thousand cities?" And, of course, you would be right. But there are a great many analogous problems (circuit design, storing and transporting millions of packages, and the like) that have the equivalent of many thousands of "cities." A saving of a few percent in the length of trucking routes, for example, can save the industry billions of dollars a year.

If a method better than trial and error for guaranteeing the most efficient (shortest, quickest, etc.) answer for even *one* of these problems is found, it will work for all of them. If, on the other hand, it can be proved that there can be no such method for even one of these problems, then there can be no such method for any of them. Most mathematicians believe that there can be no such method.

But It Looks So Easy

You no doubt remember the Pythagorean theorem:

$$a^2 + b^2 = c^2$$

Pythagoras was considering right triangles. His theorem states that the lengths of the two shorter sides determine the length of the longer side (or the other way around, which is the same thing). Specifically, the square of one short side plus the square of the other short side equals the square of the long side (the hypotenuse). In other words, $a^2 + b^2 = c^2$. If the two short sides are 3 and 4 inches, the hypotenuse is 5 (i.e., 9 + 16 = 25; the square root of 25 is 5). The lengths need not be integers (nor need they be consecutive numbers).

Now, what about a solution to the equation $a^x + b^x = c^x$ when x (the exponent—the power) is an integer *greater than two*? Is there a nontrivial solution? (A trivial solution would be that a, b, and c are all 0 and x is anything you want.)

The question of whether there is a nontrivial solution to the equation $a^x + b^x = c^x$ (x is an integer > 2; *a and b and c* are positive integers) was the most famous unsolved problem in mathematics. It is referred to as Fermat's Last Theorem, after Pierre de Fermat (1601–1665), who wrote in the margin of a book that he had "discovered a truly marvelous" proof that there could be no solution.

Fermat did not provide the proof. His margin note added that "demonstration of this proposition (is one) that this margin is too narrow to contain." Now, who would actually bother to write those words if he did not plan on someone's reading them? It is not inconceivable that Fermat was a practical joker of a particularly insidious type—after all, the man wrote in Latin—and that he knew he could flummox three and a half centuries of great mathematicians by claiming he had a proof.

In any case, there are many reasons for believing that Fermat found his "proof," if he had one, to be faulty. If someone had found just one solution for $a^x + b^x = c^x$, where x is greater than 2, that exception to Fermat's claim would, of course, have been sufficient to disprove Fermat's conjecture and to demonstrate that there could be no valid proof. No one ever found such an exception, but failure to find an exception can never prove that there is none. The exception might be the next number after the one you where you stopped checking. Here we see an asymmetry: failure to find any exception does not prove the conjecture is true, but one exception *does* prove the claim is untrue.

However, it *was* proved that there is no solution for exponents less than one million. Thus, do not bother checking to see whether, say, $16^3 + 23^3 = 47^3$, or any other combination. It does not. If there were a solution, it would have had to include numbers unimaginably much larger than those necessary to count all the particles in the universe.

Unlike science, which deals in probability-like realities, mathematics recognizes only proofs and solutions. The Fermat question can be answered only by finding a solution to the equation, which would show the theorem to be false, or *proving* that there could not be one. Incidentally, as Ian Stewart (*Game, Set, and Math: Enigmas and Conundrums*) has pointed out, there are many "close calls":

$$6^3 + 8^3 = 9^3 - 1$$

The philosopher W. V. Quine has shown that Fermat's Last Theorem can be stated in terms purely of power, rather than addition and power. This is the first of a few entries in this book that are included simply because they are nifty. An explanation would take us too far afield—out to that area where I would have no idea what I was talking about.

Fermat's theorem has finally, after three centuries, been proved, by the British mathematician Sir Andrew John Wiles. There is no solution to $a^x + b^x = c^x$ (x is an integer > 2; *a and b and c* are positive integers). Do not even ask about what the proof was. It took Professor Wiles over seven years and depended on techniques discovered in areas of mathematics unknown for centuries after Fermat. Of course, it is always *possible* that Fermat had a different, *simple* proof. But do not bet on it.

Until recently, most mathematicians merely *felt*, rather than *knew*, that there is no solution to the equation. This was certainly plausible (and, as we have seen, turned out to be true). One would think that, if there were a solution, it would show up long before the unimaginably high numbers that are the lowest that could possibly provide a solution.

Some "Fermat-like" equations do have simple solutions that are easy to find. For example:

$$a^3 + b^3 + c^3 = d^3 \text{ is solved by } 3^3 + 4^3 + 5^3 = 6^3$$

However, there are many equations that have very high numbers as first solutions, and this renders inductive thinking in mathematics *extremely* dangerous.

The Dangers of Induction

Logic and mathematics work by deduction. Each step is logically entailed in the previous step(s). Thus, if (*A*) Andy is taller than Bob, and (*B*) Bob is taller than Charles, then (*C*) Andy is taller than Charles. You do not have to measure Andy and Charles to know that Andy is taller than Charles.

Logic and mathematics do not even care whether *A* and *B* are true; their interest is only in the relationship between the premises. That is one of the things that makes math so great: you do not have to know anything. Science, while nearly always making use of logic and mathematics, works differently. It considers empirical truths ("facts") and generalizes about them. Where mathematics is certain, science is always tentative.

We have excellent reasons for believing that no cow can fly, but we always leave open the possibility that tomorrow we will spot a flying cow and all start carrying umbrellas. A science that is not tentative in this way would subordinate nature to science. And the first rule of science is that nature is never wrong, because the explanation of nature is the very purpose of science. That is why mathematics has (pretty much) *only* proof, while science can never have proof. The dangers of induction in mathematics is nicely demonstrated by Albert H. Beiler in his book *Recreations in the Theory of Numbers*.

Suppose you multiplied 2 by itself 7 times and subtracted 2 (multiplying 2 by itself 7 times is 2 to the 7th power, or 2^7):

$$2^7 - 2 = 126$$

This is evenly divisible by the exponent (i.e., 7):

$$126/7 = 18$$

Now let's try multiplying 2 by itself 5 times and subtracting 2:

$$2^5 - 2 = 30$$

This is also evenly divisible by the exponent (i.e., 5):

$$30/5 = 6$$

You might be tempted to conclude that $2^x - 2$ is always divisible by x (x stands for any integer you choose, in this last example 5). But you then try

$2^4 - 2 = 14$ (i.e., $16 - 2 = 14$), which is not evenly divisible by 4,

and

$2^6 - 2 = 62$, which is not evenly divisible by 6,

and

$2^8 - 2 = 254$, which is not evenly divisible by 8.

Aha, you say, $2^x - 2$ is always divisible by 2 only when x is an odd number (as in our first example, $2^7 - 2 = 126$). But you then try

$2^9 - 2 = 510$ and find that 510 is not evenly divisible by 9.

So where does that leave you? Well, $x = 2$ and $x = 5$ worked, and this might suggest to you that when x is a *prime number* $2^x - 2$ will be divisible by x. But when x is not a prime number, $2^x - 2$ will not be divisible by x. (Reminder: a prime number is an integer divisible without a remainder by only 1 and itself.)

At this point you are probably not sanguine about your "discovery," having been cowed by dashed hopes so many times. But now you learn that Fermat proved that $2^x - 2$ is always divisible by p when p is prime.

You will feel so proud that it will not even bother you to learn that when x *is not* prime, $2^x - 2$ is sometimes divisible by x and sometimes not. And you will be *very* glad that you did not continue using trial and error to try to find an exception to the rule that $2^x - 2$ is not divisible by p when p is not prime. You will be glad not because the rule is true (so that you could have never found an exception), but because you would not have continued trial and error long enough to find the exception and would have incorrectly concluded that $2^p - 2$ is never evenly divisible by p when p is not prime. For it would have taken you a *long* time to reach the *smallest* exception: $2^{341} - 2$ *is* evenly divisible by 341, and 341 is not a prime number. (341 equals 11 times 31.)

Incidentally, we saw that $2^p - 2$ is always divisible by p when p is prime. The 2s need not be 2s; they can be any integer. For example, $5^3 - 5 = 120$ (i.e., $125 - 5 = 120$), and 120 is evenly divisible by 3. Likewise, $9^2 - 9 = 72$, and 72 leaves no remainder when divided by 2.

Here is another example of the dangers of induction. Consider Euler's incorrect conjecture that there is no solution to $w^4 + x^4 + y^4 = z^4$. There *are* solutions to this equation. The first (i.e., smallest) was discovered by American mathematician Roger Frye:

$$95{,}800^4 + 217{,}519^4 + 414{,}560^4 = 422{,}481^4$$

(Damn, I stopped three short.)

The Lay of the Land

A mathematical system that always permits addition, multiplication, and subtraction without resulting in a new kind of number is called a "ring." For example, the mathematical system that includes only the positive and negative integers (and zero) is a ring because when you add, subtract, or multiply any numbers in the system, you get a number in the system (for example, $7 + 9 = 16$; $6 - 6 = 0$; $12–16 = -4$). However, this "ring" does not (always) permit division: 10 divided by 6 gives a fraction, and there are no fractions in the mathematical system that includes only the positive and negative integers (and zero).

A system that permits addition, subtraction, multiplication, *and* division is a called a "field." The integers alone do not constitute a field, as we have just seen.

You will remember that the rational numbers (numbers expressible as a fraction, such as 3/4, 12/12, or 4/3) together with the irrational numbers (real numbers that cannot be so expressed, such as the square root of 2) constitute the line we call the x-axis. The numbers on this line constitute a field because not just addition, subtraction, and multiplication always result in a number in the system; division also does (12 divided by 5 = 2 2/5, which is a number in this system).

The purpose of these terms is to enable mathematicians to communicate that a set of numbers has certain attributes without having to redefine the attributes relevant to the specific case. There are types of rings and fields that permit other operations without needing to go outside the system, but discussion of them would take us too far afield.

Take a Map or a Very Long Nap

Say you have an infinitely long line. This is a one-dimensional entity that is described as having one degree of freedom because the only direction you can move on the line is forward or backward (i.e., forward toward plus infinity or back toward minus infinity). Degrees of freedom are important in social statistics because each degree of freedom can represent one variable, such as like height, country of birth, and so on.

Now, say that you put a marker anywhere on the line you like. You then flip a coin. If the coin comes up heads, you move the marker one integer toward plus infinity; if tails, you move it one integer toward minus infinity. You keep tossing the coin and following the same rule. What are the chances that you will someday return to your starting point?

You may, of course, return to your starting point in just two tosses of the coin (head-tail or tail-head). But it could also be that your first million tosses are heads (unlikely as this is) and then your coin begins to act normal (coming up heads about half the time).

Even if this were to happen, sooner or later you will return to your starting place. You probably guessed this. After all, you have an infinite number of tosses, and sooner or later (probably an unimaginably long time later) you will get back to your starting place). Such is the nature of infinity.

Let's now take an infinite checkerboard instead of a line. Now we are working in two dimensions (forward/back/left/right; diagonal moves do not change anything, so we will ignore them here). Instead of a coin, we will use a spinner marked "forward," "back," "left," and "right."

Will you ever return to the square on the checkerboard where you started? Yes, indeedy. It may be in two moves or it may take a quadrillion years. But, given an infinite number of spins, you will definitely return home.

Now, let's take a three-dimensional entity (say, an infinite pile of infinite checkerboards) and a spinner with sixteen possibilities. There are sixteen because, assuming you must move to another level and not stay on the same level, you must move to one of the eight adjacent squares on the level above or one of the eight adjacent squares on the level below.

You have forever, so, once again, there is a 100 percent probability that you will return to your starting point, right? Well, actually, no.

There is a mathematical *proof* that you have only a one-third chance of *ever* returning to your starting square.

As Ian Stewart, the British mathematician and science-fiction author, points out in *Scientific American*, this means that if you are lost in the desert, you will definitely get back to your starting point at the oasis if you walk long enough (though you may be walking for a million or more years, so take *lots* of water). But if you are lost in space, you have only about a one-third chance of *ever* getting home.

Just between us, I do not understand this either. After all, you have *forever*. But when a proof conflicts with intuition and common sense, proof wins.

Happy Birthday to You

How many people have to be in a room for the odds to be greater than fifty-fifty that two of them have the same month and day birthday? One hundred? Two hundred? Surprisingly, the answer is twenty-three. (We assume that equal numbers of people are born each day, and leap year does not change the outcome.)

If there is only one person, then, obviously, the probability is 0, (There is no one to match.) If there are two people, the probability is 365/365 times 364/365 that the two do not match, and so on, until the twenty-third person arrives. Then the probability that no two have the same birthday is 365/365 times 364/365 times 363/365 and so on to 342/365. Multiply those numbers and you will find that they equal less than one-half. If the probability that no two people of the twenty-three have the same birthday, then the probability that two do have the same birthday is greater than one-half.

Johnny Carson once asked whether any member of the audience shared *his* birthday. Finding that no one did, he lightly mocked the guest who had given him the problem. Johnny misunderstood the problem; the problem did not claim that there is a fifty-fifty likelihood that one of twenty-three people will have a birthday on a specific date (in this case, Johnny's birthday), but that two of the twenty-three will have the same birthday, a much greater likelihood. In other words, picture twenty-three people being asked whether any of them has a birthday on, say, August 12. Now picture twenty-three people being asked whether any two of them have the same birthday. The latter is clearly much more likely to be the case. Indeed, the probability of one of the twenty-three having a specific birthday is one in two hundred and fifty-three, according to mathematician Warren Weaver in *Lady Luck: The Theory of Probability*.

The gist of the reasoning is this: Think of the problem as asking the probability that *no* two people share the same birthday. Now take two people. The odds of the second person having the same birthday as the first are 364 to 1. Now take three people. The odds of any two of the three having the same birthday is $364 \times 363/365 \times 365$. When you get to twenty-three people, you find that the probability of two *not* sharing a birthday is less than 0.50, so the probability that two *do* share a birthday is greater than 0.50 (i.e., more likely than not).

In general, the probability of n people *not* sharing a birthday is $364 \times 363 \times \ldots. (364 - n)/365^n$.

The Shortest Possible Completed Chess Match

P-KB4 P-K3
P-KN4 Q-R5
Black mates White.
(Martin Gardner, *Wheels, Life, and Other Mathematical Amusements*)

Why Social Scientists Do Not Do So Well

Mathematically, a "degree of freedom," a "dimension," and a "variable" are all the same thing. To say that we can move in three dimensions is equivalent to saying that our movement is a function of three variables (up-down, left-right, forward-back).

A human being is the representation of so many variables (sex, race, income, status, position in family, etc.) that it is amazing that social scientists ever find out *anything*. They work with what is effectively an "infinite dimensional manifold."

Why You Do Not See the Words "Billion" or "Trillion" in Scientific Articles

You may have noticed that scientific articles very rarely use the words "billion," "trillion," "quadrillion," and so forth. The primary reason for this is the clear superiority of scientific notation. Scientific notation puts numbers in terms of exponents. For example, a billion (1,000,000,000) is 10^9 (the 9 means 10 multiplied by itself nine times, which equals 1 followed by nine zeros); 1/1,000,000,000 is 10^{-9}; and 1, 230,000,000 is 1.23×10^{-9}. Scientific notation is *much* better when you are dealing with huge numbers.

But another reason why you never see "billion," "trillion," and so on, in scientific writing is because when the British use the words "billion," "trillion," or "quadrillion," they refer to different numbers than when Americans use the same words, and this inevitably causes confusion. If you do run into these terms, however, you need not be confused (as long as you know whether British or American usage is being employed). There is a simple way to remember the various meanings.

In British usage, the prefix indicates the number of sets of *six* zeros. Thus, a British "*bi*llion" is a 1 followed by *two* sets of *six* zeros (1,000,000,000,000), and a British "*tri*llion" is a 1 followed by *three* sets of *six* zeros (1,000,000,000,000,000,000).

In American usage, the prefix is *one less* than the number of sets of *three* zeros. Thus, an American "*bi*llion" is a 1 followed by *three* sets of *three* zeros (1,000,000,000), and an American "*tri*llion" is a 1 followed by *four* sets of *three* zeros 1,000,000,000,000).

Note that this all works for a "million" (considering "mi" means "one"). Both British and American usage call 1,000,000 a "million." This number has one set of six zeros (British usage) and two sets of three zeros (American usage).

American		**British**
Million	1,000,000	Million
Billion	1,000,000,000	Thousand million
Trillion	1,000,000,000,000	Billion
Quadrillion	1,000,000,000,000,000	Thousand billion
Quintillion	1,000,000,000,000,000,000	Trillion

In other words, an American "trillion" (four sets of three zeros) is a British "billion" (two sets of six zeros). An American "quintillion" (six sets of three zeros) is a British "trillion" (three sets of six zeros).

The difference between the systems arises because Americans consider a thousandfold increase as requiring a new level, while the British see the need for a word change only with a millionfold increase (for numbers over one million). Thus, the Americans introduce "billion" at one thousand million, while the British call this number "one thousand million" and introduce the term "billion" at the *million* million level.

One can imagine a more elegant system in which a thousand thousand is a "million," a million million is a "billion," a billion billion is a "trillion," a trillion trillion is a "quadrillion," and so on. But no nation uses this system, which is probably just as well.

We Are Stuck with Both Sexes

Say you wanted the world to have many more girls than boys. It doesn't matter *why*; this is a math problem. You might then think that the following is a great idea.

If every woman kept having babies until she had a boy, but stopped having babies as soon as she had a boy, then you would think there would be a lot more girls. Things seem like this: Only women whose first (and therefore only) child is a boy will have more boys than girls (i.e., one boy, no girls). Some women, those who have a girl and then a boy, will have the same number of boys and girls (one of each). All other women will have more girls; for example, the woman who has six girls and then a boy will have five more girls than boys (whereas no woman can have sons outnumber daughters by more than one and this is the case only with the woman who has a boy and then, as she must, stops at that point).

Well, with all those women who have a number of girls before having a boy, and with no women having more than one boy, it sure seems like there would be many more girls than boys in the world. It makes sense, but is completely wrong.

This is why. For every first child who is a girl, there will be one who is a boy. The women who have boys stop having babies. The women who have girls have a second baby. But half of these second babies are boys, so there are still as many boys as girls. The other half, the women who now have two girls, have a third baby. Half of these third babies are boys—and so forth.

Let's use numbers now. If 60 women have boys and 60 women have girls, the ratio is 1 boy to 1 girl. The 60 women who had boys stop there, while the 60 women who had girls give birth a second time. Now there are 30 new girls and 30 new boys for a total of 90 boys and 90 girls. The ratio is still 1 girl to 1 boy. The 30 women with only girls give birth a third time. Now there are 105 boys and 105 girls. And so it goes.[1]

This may make things more clear. Half of the women have a boy, but no girl. No women have a girl, but no boy. The number of the boys of women whose firstborn is a boy plus the number of boys who follow one or more girls equals the number of girls.

On the other hand, this may be the rare situation where mathematics makes things less clear than common sense. After all, half of all births are girls and half are boys. Presenting the problem in the way it has been presented serves only to introduce an element that is irrelevant.

Note

1. From Marilyn vos Savant's column, Ask Marilyn, *PARADE*, October 19, 1997.

2

Simple Does Not Mean Easy to Do

The Feel of Proof

Mathematical truths (which are always *proofs* because, as mentioned previously, there is no "probably" in pure mathematics) are truths that are certain. We *know*, because it has been proved, that we will never find a triangle on a plane in which the sum of the three angles is not 180 degrees. (There are, it must be acknowledged, some mathematicians who argue for acceptance of "extraordinarily likely" hypotheses—not merely as clues leading to proofs, but as valid on their own. Most mathematicians hate this.)

This certainty comes at a price. Unlike science, mathematics alone can never tell us anything about the real world. Mathematics *can* tell us that two apples plus two apples makes four apples, but that is because two *anything* and two *anything* makes four *anything*. Knowledge of the real world, on the other hand, must always include a fact that one could imagine not being a fact. For example, once one knows that the circumference of the earth is about 25,000 miles, *then* the mathematician can tell us that the earth's diameter is about 8,000 miles. But mathematics alone cannot tell us the diameter of the earth without a factual starting point. That takes science, with a tremendous amount of help from mathematics.

Two apples and two apples makes four apples, not because observation and experiment have shown this to be true, but because four is, in effect, *defined* as two plus two. You might well wonder, "Then how come mathematics always works in describing the real world? How come two apples and two apples *does* make four apples?" Nearly every great thinker has also wondered this, but no one has yet come up with an answer that does much more than restate the question.

Thus, take the truth that "we'll never find a bachelor with a wife." The scientist can dismiss this as a trivial truth because it is a "tautology," a truth that is true *by definition*. The mathematician can, of course, do the same. But his dismissal is not on the grounds that it is a tautology, but because it is an *obvious* tautology.

If the mere fact that something is a tautology were to make it trivial, then all mathematics would be trivial because every mathematical proof is a tautology: its conclusion is entailed in the definitions of its premises (e.g., a few axioms and the properties of numbers). Just as we will never find a bachelor who is married, we will never find a triangle on a plane whose angles do not add up to 180 degrees. In each case, the underlying definitions *entail* the finding. Put another way, a God who is infinitely intelligent but not able to predict the future would know all mathematics instantly but not all that will happen in the future of the real world. (This assumes that reality is random at bottom, which it seems to be.)

Because mathematics is ultimately based on definitions and logic (that is what permits the certainty that is precluded by the real world), it also avoids the vagueness inherent in words. Consider a question such as, "Do chimps have 'language'?" Superficially, this seems to be a fascinating question. A bit of thought, however, makes it clear that this question is not amenable to a satisfying answer because "language" cannot be sufficiently and rigorously defined to permit an answer. If "language" means "warning a member of your species," then beavers have language. (They slap their tails or something.) If "language" means "writing a good sonnet," then most of us do not have language. The interesting scientific question is not "Do chimps have language?" but "What is it that chimps do to communicate?"

Linguistic vagueness is inescapable because words are categories describing a world not built of categories. Even such seemingly iron-clad categories as "living" and "nonliving" are ultimately incapable of immunization against the continuity of reality. "Living" and "nonliving" *seem* easily distinguishable, as indeed is the case if you are comparing a person and a rock. But when you try to give the definitional dividing line, it becomes clear that it is not so easy. Most people will conclude that "living" is defined by "reproducing itself" and will be most annoyed to find that they have defined crystals as "living." The problem devolves from attempting to capture the virtually continuous nature of nature (at least at the supra-atomic level) with the discrete nature of words.

Here is another example. When I was in school, biologists argued about whether the euglena (one of those little guys you see through a microscope) is a plant or an animal. It has some of the characteristics that define a plant and some that define an animal. This problem of classification did not stop nature from making more euglenas. Today, a new taxonomy has overcome the euglena problem, but no one doubts that we will soon discover a horse with wheels or some other creation that nature whips up to keep us in our taxonomic place.

Thus, all attempts to categorize nature must fail. A new taxonomic system may be without failure momentarily, but nature always eventually creates the uncategorizable. And even sooner, we can always *imagine* an exception and can specify its properties.

Except in mathematics. Because mathematics defines its universe and is not beholden to nature, only in mathematics can we avoid the taxonomic problem. So, for example, a number is 1 (or 12 or 1,247,351 or pi or whatever) or it is not 1 (or 12 or 1,247,351 or pi or whatever). It cannot be "sort of 1."

There is, it must be acknowledged, a tiny group of mathematicians who deny "the law of the excluded middle" and claim that we cannot say that a number is 1 or not 1. They are not much fun at parties.

An Embarrassing Admission

Okay, let's get this out of the way now and save the mathematicians the awful anticipation of waiting for exposure of this terrible embarrassment. Yes, it is true; they cannot define division by zero. When you or I screw up, we call it a "mistake." When a scientist screws up, he calls it an "anomaly." When a Ford executive screws up, he calls it an "Edsel." When a mathematician screws up, he calls it "undefined."

Consider what it means to say that "six divided by two is three." It means that when you multiply two by three you get six. What would it mean to say that "six divided by zero is x"? It would mean that zero times x is six. But there is no number that, when multiplied by zero, equals six. Zero times any number is zero. You can see the problem and the need to call division by zero "undefined."

When we talk about zero, we are not just talking about any old number (such as 43 or 6,412). We are talking about the beginning of all numbers, the godhead of the mathematical universe. Indeed, once the logician has zero, the name of an "empty set," he can construct the whole of mathematics.

We in the other academic departments often try to run interference for the mathematicians to save them embarrassment. When we are asked whether "those mathematicians have defined division by zero yet," we answer that the mathematicians "have had the flu; we're sure that they will define zero as soon as they recover." But no one is buying this anymore.

The Birthplace of All Numbers

Note that the mathematician's declaring zero, the birthplace of all numbers, to be functionally undefined resonates with the physicist's declaring the big bang, the birthplace of the physical universe, to be a singularity. Perhaps at the core of things there can be only mystery. Some would say that this is an expression of the mystery that must remain, even if we were to learn all that can be learned. Others, those with a religious sensibility, would say it is the door to God's house.

If you have a terrific memory (or were recently in high school), you may remember learning that the different types of numbers we are most familiar with can be graphed on a straight line. First there are the natural numbers (positive integers such as 1, 2, 3, 4 . . .). Then there are the negative integers (such as the negative $1,200 in your bank account) and 0. Then we can fill in the spaces with fractions that end (such as 1/2, which is 0.5) or do not end (such as 1/3, which is 0.3333333 . . ., or 2/7, which is .285714285714285714285714 . . .) but always have repeating decimal patterns. These are called "rational numbers."

This still leaves some spaces on the number line because the rational numbers (the numbers expressible as a fraction of two integers) have between them an infinity of numbers. Between any two fractions there is an infinity of irrational numbers, numbers *not* expressible as a fraction composed of two integers. These space fillers are the "irrational numbers."

The number that is the square root of 2 is an irrational number. You can create fractions that, when squared, get closer and closer to 2, but they are always a tiny bit below or above 2. Thus, the number that, when squared, gives 2 is not a fraction; it is "the square root of two." "The square root of two" is the name of this number, just like "3" is the name of "three."

Then came the "imaginary numbers," multiples of the square root of minus one. At first this scared a lot of mathematicians. There seemed good reason to be scared. The most obvious of these is that a positive number times itself had always given a positive number ($1 \times 1 = 1$ and $4 \times 4 = 16$). Zero times itself gives zero. And a negative number times itself also gives a *positive* number ($-1 \times -1 = 1$; $-4 \times -4 = 16$). So what kind of number could it be that, when multiplied by itself, gives a negative number?

Well, the mathematicians finally decided it could be an "imaginary number." After all, if a new kind of number could not break *any* of the rules of the old kinds of numbers, then we could not even have had the fractions. Indeed, if a supposed new kind of number did not break any of the rules of the old kinds of numbers, it would not be a new kind of number.

Thus, the unfortunately named "imaginary" numbers are every bit as legitimate as the numbers known long before; the square root of negative one (called "*i*") is just as much a number as six or ninety-seven. Such numbers are not any more imaginary than any other numbers. One is tempted to say that they are just as "real," but, unfortunately, the rational and irrational numbers, taken together, are already named "real." That is why this new kind of number is called "imaginary."

As long as a new type of number follows consistent rules, it is legitimate. Consistency in mathematics means that there is no contradiction. If there is a contradiction, a *single* contradiction, the entire structure collapses. (A system with a contradiction permits one to prove anything.) This is the downside of simplicity and elegance and why mathematics takes such guts. You do not know humiliation until you have published an article in a mathematics journal that has a sophisticated version of $1 + 1 = 7$ for all to see.

Incidentally, this "as long as it works and does not introduce inconsistency" is not always easy to swallow. In high school, we learned of Zeno's paradox, which states, for example, that when you walk toward a wall, you must cover half the distance before reaching the wall, then half that distance, then half that distance, and so on. So you will never reach the wall. But you *do* reach the wall.

Now, while there are much more sophisticated answers for why you reach the wall, even though it seems you always have some distance to go, the answer we got in high school was that an infinite series (the steps) adds up to a finite amount (the starting distance to the wall). No proof was given for this claim, and it struck me that it had a certain ad hoc quality, that it was whipped up to "explain" why you reached the wall without really explaining things, but only redefining them.

Likewise, we learned that the unending decimal, 0.999 . . . is the same thing as 1.0. "Says who," I wondered.

And again, it was claimed that the infinity of even numbers is the same size as the infinity of integers. Now, you would think that the infinity of integers, which includes both the odd and even numbers, is twice as large as the infinity of even numbers. But, no. Size is defined by a matching process (1 matched with 2, 2 matched with 4, etc.). And because you cannot run out of numbers in an infinite series, matching the integers and the even integers is possible, so the infinity of even numbers is as large as, if less dense than, the infinity of all integers. (There *are* larger infinities—an infinite number of them—that the even numbers, or even all the integers, cannot match. But we will ignore that here.)

This too struck me at first as a bit dubious, an evasive redefinition of things that seemed rather shoddy. But, as indicated, my dubiosity was not justified because all of these practices, of which I had unfairly been skeptical , enlarged mathematical knowledge without introducing any incorrectness or contradiction, just as had been the case with new types of numbers, such as fractions and irrationals.

One Hundred Hamburgers + Red and White Balls to the Max

Now, let us go back to types of numbers. There are many other types of numbers. Next on the ladder is "complex numbers," which combine real and imaginary numbers. The list goes on until it reaches a level of abstraction that very, very few have observed.

While descriptions of the other types of numbers would take us too far afield even if I knew enough about these numbers to describe them, it is worth making a point about "infinity" because the very properties that make infinities new types of numbers seem so unintuitive.

If you have a hundred hamburgers and a hundred and one people to eat them, the hamburgers and the people will not match up; there will be one person who does not get a hamburger. This is pretty obvious. But the reason the hamburgers cannot be matched up one-to-one with the people is that we are dealing with finite numbers and there are not enough hamburgers. Similarly, if a box holds fifty white balls and fifty red balls and you take out a white ball, the odds (slightly) favor the next ball's being red. There is an end to the number of white balls. So when the first ball you choose is white, there are more red balls left. This principle would, of course, obtain if the first ball were red.

However, what if we have a box with an infinity of red and white balls? Because infinities never get to an end, taking out a white ball does not imply anything about the color of the next ball. Failure to understand this has led to *a lot* of people losing their money. The fact that red comes up twelve times in a row on an unbiased roulette wheel does not change the odds from even money on the thirteenth spin of the wheel, which is also the case with the box with an infinity of balls.

It is obvious that the infinity of white balls is the same size as the infinity of red balls. It is obvious in the same sense that twelve hamburgers and twelve hamburger eaters are groups of the same size: they can be matched one-to-one.

Consider, however, two infinities, one of which seems obviously bigger than the other. We might choose *A*, the infinite set of integers (1, 2, 3, 4 . . .), and *B*, the infinite set of even numbers (2, 4, 6, 8 . . .). Common sense tells us that the first set is bigger—twice as big. But remember that the way we measure the size of a group, finite or infinite, is to match them one-to-one. Can groups *A* and *B* be matched?

Yes. In group *A*, 1 is matched with 2 from group *B*. Then 2 in group *A* is matched with 4 from group *B*—and so on forever. Group *A* may in some way be more densely packed, but the two infinities are the same size. Infinities of this type are called "countable" because there is an integer to match up to each element. (Note that the infinity of positive and negative integers is the same size as the infinity of just the positive integers.)

Despite the fact that we have seen that infinities can have counterintuitive properties, it seems obvious that all infinities are the same size. In other words, you just match each element in any infinity with an integer and you have shown that all infinities are the same size.

Cantor's Diagonal Proof and Casti's Way

Note, however, as Georg Cantor did a hundred years ago in one of the greatest of all mathematical insights, that it is not self-evident that you *can* always match every element of an infinity with an integer. Here is Cantor's wonderful proof that you *cannot* always do so.

Let us list all of the numbers—including unending decimals—between 0 and 1. We will treat numbers that are not unending by their unending equivalents (i.e., 1 is 1.00000 . . . . Now, we cannot in reality make an infinite list, but we can list a few numbers and demonstrate that, however long the list of numbers, there will always be a number not included in the list. By doing this, Cantor proved that there is a larger infinity than the ones that are countable. His proof—called Cantor's Diagonal Proof—gives new meaning to "elegant."

Cantor made such a list. It does not matter what order the numbers are in, so let's arbitrarily choose numbers between zero and one. (The same point obtains whatever range one chooses, as all countable numbers are countable with the countable numbers between zero and one.)

0.763498276 . . .
0.500000000 . . .
0.267037143 . . .

0.123456789 . . .
0.987654321 . . .
0.555555555 . . .
0.273403949 . . .
etc.

Now, draw a diagonal line, beginning with the first digit of the first number, the second digit of the second number, the third digit of the third number, ad infinitum. The unending diagonal number we get will begin 7074553 . . . .

Add 1 to each digit of our diagonal number. The new number will begin 8185664 . . . . Notice that the diagonal number we end up with cannot be the same as the first number because its first digit is an 8, not a 7. It cannot be the same as the second number because its second digit is a 7, not a 6. It cannot be the same as the third number . . . well, you get the idea. The infinity of listed numbers does not include the diagonal number. Therefore, the diagonal number is a member of an infinity that cannot be counted (i.e., an infinity larger than the countable infinities). This demonstrates that some infinities are larger than others. Moreover, because the same sort of argument can be made against *any* infinity claiming to be the largest, there is no more a largest infinity than there is a largest number.

The mathematician and science author John L. Casti, has a marvelous way of demonstrating the diagonal proof that is easier for some people to see:

Consider these six names:

Twain
fUrman
beRry
sprIng
lockNer
herzoG

Create a word using the letter after the first letter of the first word, the letter after the second letter of the second word, and so on. You will get "turing." This must be different from every word on the list because it will differ from the first word by a different first letter, the second word by a different second letter, and so on.

I hope you now have some slight feeling for what elegance is. This book is a compendium of examples of elegance, with some other stuff thrown in just because it is interesting. The point is not to explain in any detail, but to give the reader a taste that will, I hope, lead to exploration of one of the many paths hinted at in these pages.

I cannot stress this final point enough: This book will not teach you mathematics. I am far, far from qualified to do this, and there are many, many mathematicians—some of them, unfortunately, unemployed—who are eminently qualified to do so.

My purpose is simply to provide a sort of tasting menu to whet your appetite. I attempt to serve up fascinating and beautiful findings of mathematics, in the hope of persuading you to follow up on those you find most appealing by searching out the works of mathematicians who have described those findings in detail.

A Million

Picture a die, like one used for gambling only much smaller, just one-tenth of an inch on each side. A million of these lined up horizontally will stretch out about one and two-thirds miles. Ignoring the height and width of the dice, this is a one-dimensional arrangement.

Now, let's arrange the dice flat in a square, a thousand dice long and a thousand dice wide. The square holding the million dice will be a bit over eight feet (a hundred inches—one-tenth inch times one thousand) on each side. Ignoring the third dimension, the height of the die, this is a two-dimensional arrangement.

Finally, let's make a square of a hundred dice by a hundred dice (i.e., $100 \times 100 = 10{,}000$ dice). Now make ninety-nine more such squares and pile each on top of the previous one. You now have a cube, $100 \times 100 \times 100 = 1{,}000{,}000$. This cube will be only ten inches by ten inches by ten inches. This is a three-dimensional arrangement, probably much smaller than you would have guessed.

This Kid Shows Promise

When Carl Gauss, the greatest of all mathematicians in some people's view, was seven years old, his teacher did not feel like lecturing to his second graders one day. So he gave the boys a math problem that would easily take them the entire class: add the numbers from 1 to 100. Seconds later, young Carl raised his hand and told the teacher that the sum was 5,050. Both the teacher and the other students were dumbfounded.

Carl had realized that he could add pairs from the two ends: that is, $(1 + 100) + (2 + 99) + (3 + 98) \ldots$. Each pair of numbers adds up to 101. Because there are fifty pairs of numbers, the total is 50×101, or 5,050.

A Mathematical Fact

Mathematics is full of facts that you would never guess are facts. For example, Joseph Louis Lagrange proved that every natural number (i.e., positive integer) is equal to the sum of four or fewer square numbers: for example, $23 = 9 + 9 + 4 + 1 = 3^2 + 3^2 + 2^2 + 1^2$. Trial and error establishes that not all natural numbers can be represented as the sum of *fewer* than four squares, though some can: for example, $18 = 3^2 + 3^2$.

You Will Never Get Rich

Most people know that the more often interest is compounded, the more money you end up with. Banks know that people know this and compete by offering ever-more-frequent compounding. What banks also know is that the difference between semiannual compounding and daily, or even second-by-second, compounding is insignificant.

Let's say you have $100 in a savings account and the bank compounds the interest *annually* (i.e., *once* a year). Your money will double (i.e., grow to $200) in, for example, ten years. Now, if the bank compounds the interest *semiannually*, your $100 will grow to about $269, rather than $200, in the same amount of time.

This might lead you to the conclusion that ever-more-frequent compounding will engender similar increases. Banks occasionally exploit this belief by compounding ever-more-frequently.

But that conclusion is not correct. In fact, even if the $100 dollars is compounded second-by-second, it will grow to only about $272 (actually a tad less) in the same time that semiannual compounding would make it grow to $269.

This is because compounding approaches the limit of the "*e*" (2.7182 . . .). Why? You might well ask *why* does *e* equal 2.7182 . . . but you will not be satisfied with the answer, which is "because it does." The *e* is an irrational number—a number not expressible as a fraction composed of two integers—that is, as we have seen was the case with the square root of 2, just as legitimate a number as 6 or 3/4. The *e* is one of those numbers, like π, that is a mathematical constant that keeps popping up in divergent mathematical areas for no obvious reason.

Each increase in the number of times your savings is compounded *does* increase the total; daily compounding is better than weekly compounding. But very soon the increments become so small as to be insignificant.

The reason that we are surprised to find that second-by-second compounding does not make us far richer than monthly compounding is that we tend to think in terms of what is known as *linear* functions. If you make ten dollars an hour and work for four hours, you get forty dollars. If you work for eight hours, you get eighty dollars, twice as much as for four hours. These are linear functions; the rates of increase are what you intuitively expect.

Many functions, however, are nonlinear and behave in unexpected ways. Compound interest is nonlinear. While there is always *some* increase when you increase the number of times you compound in a given period, the increases get smaller and smaller very quickly.

3

Gamesmanship

Great Questions

The constraints I have set on this book would preclude my discussing these questions even if it were not the case that they far exceed my range of vision. The reader is encouraged to consult the works that have been written on each of these. Volumes have been written, yet the questions are still unanswered. That is why they are great questions. The first four questions are purely and *arguably* the four greatest unanswered mathematical questions. The others are scientific (empirical) questions that have strong mathematical components (though question 5 is difficult to categorize).

1. What is the most simple algorithm that identifies the primes? This question is inextricably entwined with Riemann's Prime Number Hypothesis, a speculation about the distribution of primes that is a wonderful next step for the reader with a lively curiosity about mathematics. (This is the question I would ask God if I could ask just one question.)
2. P = NP (the NP completeness problem)?
3. Is there an infinity larger than the rational numbers but smaller than the real numbers (the continuum hypothesis)?
4. Is there an even number that is not the sum of two primes (Goldbach's conjecture)?
5. Why does mathematics work in the real world?
6. What is the topology of the universe?

7a. Will the universe close?

7b. If the universe closes, will time go backward, and if it does, what does this mean?

8. Why does time have direction?
9. Is there a way around the antirealism implied by Bell's Theorem and the relevant experiments? If not, which assumption should we give up?
10. What mediates between the small (quantum world) and the large (classical world)? In other words, what is the state of Schrödinger's cat?
11. What goes on inside a black hole? Does it have an "other side"?
12. Does it rain in Indianapolis in the summertime?

Play Chess So That You Never Lose

You already know that there is a way to play tic-tac-toe that guarantees you never lose. When you were very young, you probably found tic-tac-toe to be a challenging game. But you soon discovered a strategy that permitted you to avoid losing; the strategy enabled you to tie the game at worst, and to win if your opponent made a bad move.

You accomplished this intuitively, but you could have done so by making a matrix in which you wrote down every move your opponent could make, every response you could make to your opponent's move, every response your opponent could make to your response, etc. By simply following a path that ended with a tie you would guarantee a tie even if your opponent always made his best possible move.

All of this is true of chess (and checkers). It has been proven that there is a guaranteed nonlosing strategy for every game that has complete information (i.e., each player knows the state of the game at each moment, and there are no hidden cards, as there are in poker) and in which chance plays no role (e.g., games using dice). Such games guarantee a draw (tic-tac-toe), a victory for the player who goes first (some versions of nim), or a victory for the player who goes second (other versions of nim). So, you might ask, why do they have chess championships?

Because, while we know that there is a correct way to play chess (i.e., a strategy that avoids losing), we have virtually no idea what that strategy is. (It is often the case in mathematics that it can be proven that there is an answer to a question without anyone having any idea what the answer is.) In chess, we do not know whether the strategy results in a draw, a victory for the player who goes first, or a victory for the player who goes second.

The reason we do not know this is that, while the total number of possible states of the "board" in tic-tac-toe is relatively small, the number of possible states in chess is, while finite, approximately 10^{118}, a number unimaginably much greater than that required to count all of the electrons in the universe. Indeed, Seth Lloyd, professor of mechanical engineering at Massachusetts Institute of Technology, has estimated that the number of states the universe has ever been in—that is, the number of physical events that have taken place, including every change in the state of every particle—is only 10^{120}.

However unknowable the master strategy for chess, a master strategy for tic-tac-toe is easily stated.

Play Tic-Tac-Toe and Never Lose

As a child, nearly everyone learns how to play tic-tac-toe without losing. However, it is not so easy to articulate the strategy one has intuited. This is the best Mike Mayers—a dab hand at numbers (and my brother-in-law)—and I can do. Elegant it is not, but it is the best we can do. See if you can remove one of these rules (or part of one) and still guarantee a draw. Following these rules guarantees that you will not lose (i.e., you will draw), even if your opponent plays a perfect game (and sometimes win if your opponent does not follow the strategy). It does not matter whether you go first or second.

Note that these rules do not guarantee you will win every game in which a victory is possible, only that you will never lose. An opponent who randomly selects moves or who plays badly will often give you an opportunity to win that will require that you make moves not given by these rules, while following these rules will result in a draw. However, a set of rules that guarantees victory in such cases would be far too unwieldy to be of use, and the correct moves in such cases will be intuitively obvious.

The *first* rule that applies determines your move:

1. On your first turn, go in the center if you can, otherwise go in a corner.
2. If you have only the center square and your opponent has (only) two diagonally opposing squares, go in a noncorner.
3. If any line has one empty square and two squares with your opponent's mark, go in the empty square.
4. Unless there is no choice, do not complete a line that already has one X and one O.
5. At any point that you cannot follow any of these rules, go anywhere.

Even Fermat Can Get It Wrong

Fermat believed that there was no integer solution to the equation $a^4 + b^4 + c^4 = d^4$. This is not surprising as the smallest solution is believed to be $95.000^4 + 217{,}519^4 + 414{,}560^4 = 422{,}560^4$. This is one of many lessons demonstrating how induction can never be proof in mathematics because the next numbers might be the proof.

Hard as It Is to Believe

John Allen Paulos, author and math professor at Temple University, gives an interesting example of a counterintuitive reality.

Say that Babe Ruth has a higher batting average for the first half of the season than Lou Gehrig. Say also that Ruth has a higher batting average for the second half of the season.

Ruth *must* have a higher batting average for the season, right? Wrong. Say that for the first half of the season Ruth had 55 hits in 160 at bats for a 0.344 average and that Gehrig went 82 for 240 for 0.341. And say that for the second half of the season, Ruth went 60 in 240 for 0.250 and Gehrig went 38 in 160 for 0.238. Ruth has a higher average for each half of the season. But for the season, Ruth went 115 in 400 for 0.288, while Gehrig went 120 in 400 for 0.300.

Note that the players had the same number of at bats (i.e., the same number of chances) for the season. Had they had the same number of at bats in each half of the season, it would have been impossible for a player to lead for each half and yet come in second for the season.

To see all of this a bit more easily, here is another example. Say that for the first half of the season Don Drysdale has 9 victories and 7 losses for a winning percentage of 0.563 and that Sandy Koufax is 3 and 3 for 0.500. Say that for the second half of the season Drysdale is 3 and 0 for 1.000, and Koufax is 10–3 for 0.764. Each pitcher has 19 decisions for the season. Drysdale has a better winning percentage in each half of the season, but Koufax has a better percentage for the whole season. (Drysdale is 17–7 for 0.632; Koufax is 13–6 for 0.684.)

Not Everything Has Its Limits

Consider an infinite series of numbers, say 1 + 2 + 3 + 4 . . . . Clearly this series has no limit; as you add numbers, the series gets bigger, without limit. Whatever number you choose—even one with a quadrillion digits—the series will eventually pass that number, and the sum of the series will forever get larger.

Now consider this series: 1/2 + 1/4 + 1/8 + 1/16 . . . . This series does have a limit: 1. You can forever get closer and closer (and, taking the series as a whole, consider the total to be 1), but you can never get a total past 1.

It is not always easy to tell from inspection whether a series has a limit. As Martin Gardner points out, consider the ubiquitous "harmonic series," 1/2 + 1/3 + 1/4 + 1/8 . . . . The terms become increasingly smaller, leading one to suspect that the series (like 1/2 + 1/4 + 1/8 + 1/16) approaches a limit. It certainly seems to.

It takes this pokey series 12,367 terms to pass 10. Worse yet, the number of terms it takes to pass 100 is more than 1,000,000,000,000, 000,000,000,000,000,000,000. But the series has no limit. Pokey or not, it just keeps on going its pokey way.

Nim

Nim is a game that has been played in various forms on at least four continents for at least four centuries. Like tic-tac-toe, it is a challenging game until one realizes that there is a correct way to play.

In the case of tic-tac-toe, there is a correct way for both players, and if both players make the correct moves, the game will always end in a tie.

In the case of nim, when one player makes the correct moves, he will always win. Whether this is the player who goes first or the player who goes second depends on the variation of nim being played.

Where the perfect strategy for tic-tac-toe is discoverable by a bright child, discovery of the correct nim strategy takes a mathematical intuition of the highest order for one without mathematical experience.

Nim is often called "the Marienbad game" because it was played, in its most familiar version and the one given here, in the movie *Last Year at Marienbad*. In the movie, and traditionally, matches are used, but pencil marks are far more convenient.

The correct nim strategy is given on page 31. Try playing a few games before looking at the correct strategy. Once you do, you can never again enjoy nim as a challenging game. On the other hand, in learning the strategy you will experience a deep and, for the mathematically inclined, exciting glimpse of the deep connections underlying the game.

Notice how all this nicely sums up both the gains and losses that come with modernization. The discovery of the correct strategy for nim was a solution to a problem that had gone unsolved for centuries. At the same time, the solution meant the loss of a game that had provided pleasure and social contact for millions and had enabled thousands to make a living.

How to Play Nim

Draw the following layout with a pencil.

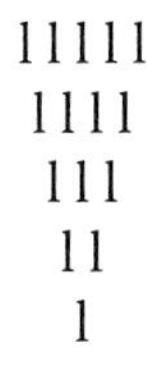

Two players alternate turns. On each turn, a player crosses out marks *on one line only*. He may cross out as few or many as he wishes, but he must cross out at least one. The player who is forced to cross out the last mark loses.

Notice that position on a line does not matter. For example, if the player going first wishes to take away two marks from line three, it makes no difference whether he takes away the two on the right of line three or the two on the left of line three. Thus, for ease of play, it is traditional to take away marks from the right. So this player would cross out the marks indicated by x's.

```
 1 1 1 1 1
  1 1 1 1
   1 x x
    1 1
     1
```

Here are a few layouts to get you started.

```
1 1 1 1 1 1 1 1 1 1 1 1 1 1 1
   1 1 1 1 1 1 1 1 1 1 1 1
      1 1 1 1 1 1 1 1 1
         1 1 1 1 1 1
            1 1 1

1 1 1 1 1 1 1 1 1 1 1 1 1 1 1
   1 1 1 1 1 1 1 1 1 1 1 1
      1 1 1 1 1 1 1 1 1
         1 1 1 1 1 1
            1 1 1

1 1 1 1 1 1 1 1 1 1 1 1 1 1 1
   1 1 1 1 1 1 1 1 1 1 1 1
      1 1 1 1 1 1 1 1 1
         1 1 1 1 1 1
            1 1 1

1 1 1 1 1 1 1 1 1 1 1 1 1 1 1
   1 1 1 1 1 1 1 1 1 1 1 1
      1 1 1 1 1 1 1 1 1
         1 1 1 1 1 1
            1 1 1
```

The Nim Winning Strategy

Read "How to Play Nim," on page 29, before reading this section.

This method works for all forms of nim. The 5-4-3-2-1 layout used in this example (the most commonly used layout) always results in a win for the player who goes first, assuming that he follows the strategy given.

In the case of some alternative layouts (for example 7-5-3-1) or rules (for example, the person who takes the last match *wins*), the player who goes second always wins. The strategy remains the same, simply be certain to let your opponent have the first move. (There is nothing he can do to avoid losing.)

Here we also assume that the goal is to force your opponent to cross out the last mark (i.e., the player who crosses out the last mark *loses*). This is the more commonly used variation. If you are playing a variation in which the player who crosses out the last mark *wins*, simply follow the strategy given, but play to "lose."

Count the number of marks in each row and memorize the number associated with the number of matches in a line.

Layout = Number
l l l l l = 101
l l l l = 100
l l l = 11
l l = 10
l = 1

Whenever there are three marks in a row, that row equals eleven. Thus, if the first player crosses out two marks from the top row, the value of the top row goes from 101 to 11. (For the moment, do not worry about *why* five matches equals 101, four matches equal 100, etc.)

If you forget the code, you can reconstruct it easily. Begin with one match or mark, which equals 1. Two matches or marks will equal the next (ten-based) number that can be made of just 0s and 1s, which is 10. Three matches or marks equals the next (ten-based) number that can be made of just 0s and 1s, which is 11. Continue this to 100 and 101.

Always leave your opponent with a total of *all even digits* (not simply an even number; i.e., 210 is not a good total to leave your opponent). However, when at the end of a game you find that you must leave your opponent with all remaining rows containing only one digit each, leave him with an *odd* total). This will always be possible.

Assume you go first. Consider your first move. The total of a full layout is 223 (101 + 100 + 11 + 10 + 1). You want to leave your opponent with 222. (Any other possible total will have an odd digit.) You can cross out one mark in the top row (making 101 become 100), one mark in the middle row (making 11 become 10), or the mark in the bottom row (removing 1). You cannot remove, say, one from the second row (making 100 become 1) because the total will then be 124, which has an odd digit.

To play the game and always win, you need only remember or reconstruct the code (lllll = 101; llll = 100; lll = 11; ll = 10; l = 1) by the 1-10-11-100-101 method described. To see deeply into the structure and beauty of the mathematics, study closely the derivation of the code.

Matches	Number of Matches in 10-Based System	Number of Matches in Binary System	Binary Number Translated into 10-Based Number
lllll	5	$\underline{1} \times 2^2 + \underline{0} \times 2 + \underline{1} \times 1$	101
llll	4	$\underline{1} \times 2^2 + \underline{0} \times 2 + \underline{0} \times 1$	100
lll	3	$\underline{1} \times 2^1 + \underline{1} \times 1$	11
ll	2	$\underline{1} \times 2^1 + \underline{0} \times 1$ 10	10
l	1	$\underline{1} \times 1$ 1	1

Five Great Women Mathematicians

The five greatest women mathematicians are arguably, in chronological order, Hypatia, Maria Agnesi, Sophie Germaine, Sonya Kovalevsky, and Emmy Noether, clearly the greatest female mathematician. Only Kovalesky ever married, and hers was a platonic marriage of convenience. Make of this what you will, but it is unlikely that it represents a coincidence.

4

Safe Landings

Do Not Bother Looking for the Largest Prime

You will remember that a prime number is an integer that is evenly divisible only by 1 and itself. The numbers 2 and 3 are prime; 4 (which is divisible by 2) is not. The only even prime is 2, as all other even numbers are evenly divisible by 1, themselves, *and* 2. (The number 1 is usually not considered a prime, for technical reasons.)

Euclid (or one of his cronies) wondered whether there was a finite number of primes or an infinite number of them. In other words, with reference to all primes—2, 3, 5, 7, 11, 13, 17 . . . p—is there always a prime larger than p? Euclid's proof is the quintessence of mathematical simplicity and beauty—in other words, elegance.

Let us assume that the primes are finite, so there is a largest prime. We will call that prime p. Multiply all the primes up to and including p (i.e., $2 \times 3 \times 5 \times 7 \ldots p$). Call the product Q. (You could instead multiply all the integers—i.e., $1 \times 2 \times 3 \times 4 \times 5 \times 6 \times 7 \ldots p$—but this amounts to the same thing because the nonprime numbers reduce to primes.)

Now add 1 to Q.

$Q + 1$ is not evenly divisible by 2 or 3 or 5 or 7 . . . p or any combination of these (i.e., division by any of these leaves a remainder).

If $Q + 1$ is composite (i.e., not prime), it is evenly divisible by a prime *larger* than p (not p or any prime smaller than p because these leave a remainder). Therefore, p is not the largest prime.

If, on the other hand, $Q + 1$ is prime, then it is a prime larger than p.

In either case, there is a prime larger than p.

Should $Q + 1$ get cocky and think that it is the largest prime, we need merely give $Q + 1$ the role previously played by p and hire S to play Q's role. (Ad infinitum.)

Consider this example. Say we make $p = 3$. Then $Q = 2 \times 3 = 6$ and $Q + 1 = 7$, a prime number larger than 3.

If, on the other hand, we make $p = 13$ (i.e., $2 \times 3 \times 5 \times 7 \times 11 \times 13$), then $Q = 30{,}030$. And $Q + 1 = 30{,}031$, which is not prime and must require a prime larger than p.

Because 30,031 is a small number, it is known that its prime divisors are 59 and 509. You learn this from a table of prime numbers. For really large numbers, it takes centuries or much, much longer to determine their primes. That is what makes "unbreakable" codes unbreakable. In practice, if you had billions and billions or more of years, you could solve any practical code.

In other words, this works if, instead of 3 or 13, we made p equal, say, a skillion, skillion, skillion (or any other integer). The logic is the same. (Actually, there is no number "skillion." It just means "any really, really big number," as big as you want.)

If you try this, say, a trillion times, you will find that most of the $1 + (2 \times 3 \times 5 \ldots)$ are composite, not prime.

These Euclidians—they were really, really good.

What?

Picture a line containing the integers $(\ldots -3, -2, -1, 0, 1, 2, 3 \ldots)$. Starting at zero, toss a coin. If it lands heads, go right; if tails, go left. Do this forever. Common sense says that you can choose a number and that, sooner or later, you will land on it. Common sense is correct; on a one-dimensional surface, a line, you will, having forever, return to your starting point.

Do the same thing on a plane, using a spin wheel divided into four equal sections, 1, 2, 3, and 4. Section 1 is left, 2 is right, 3 is forward, and 4 is backward. Common sense says that you can choose a number and that, sooner or later, you will land on it. Common sense is correct; on a one-dimensional surface, a line, you will, having forever, return to your starting point.

Now, do the same thing in three dimensions, say, three checkerboards on top of one another. Use a spin wheel divided into the eighteen possible moves (i.e., straight up, straight down, up and left, etc.; movement on the same plane is not permitted because it confuses things, but it does not change them in any relevant way).

Common sense says that, sooner or later, you will come back to where you started. Common sense is incorrect. There is nearly a two-thirds chance you will never land on your original spot.

I do not understand this either. But it is true.

Is Every Even Number Larger Than the Sum of Two Primes?

A prime, you will remember, is an integer evenly divisible only by 1 and itself. It is obvious that 2 can be the only even prime (because all other even numbers are divisible by 1, themselves, and 2).

Now, aside from having a quintessentially ecumenical-sounding name, the Prussian mathematician and historian Christian Goldbach, in a letter to the great Euler, is noted for one thing: in 1742, he conjectured that every even number is the sum of two odd primes or, in the case of 2, 1 + 1 (e.g., 2 = 1 + 1; 4 = 2 + 2; 6 = 3 + 3; 8 = 5 + 3; 10 = 5 + 5; 12 = 7 + 5, . . . 98 = 19 + 79 . . .).[1]

Is Goldbach's simple conjecture true? Two and a half centuries later, we still do not know. We do know that every even number is the sum of not more than 300,000 primes.[2] So all that need be done is to shave the 300,000 down to 2.We know that there is some number above which any number is the sum of not more than four primes. The problem is that no one knows what that number is. It is not 37.

We also know that every even number through 100,000,000 is the sum of two primes, but there are a lot more than 100,000,000 even numbers larger than 100,000,000. Indeed, the even numbers are unending.

In science, a hundred million cases without an exception would justify the strongest belief a scientist can hold, one that is always tentative (the next empirical case may be an exception), but one that permits science to proceed.

In mathematics, all that counts is proof. Nothing less than a proof that every even number is the sum of two primes (or an exception that would refute the conjecture) counts for anything.

I Knew the Answer Was Either Graham's Number or 6

According to *The Guinness Book of World Records*, the largest number ever used in a mathematical proof was "Graham's Number" This number required the invention of a new notation system because, even using towers of exponents, there would not be enough room in the universe to express the number. Graham's Number arises in a certain problem in combinatorics, the branch of mathematics that tends to produce the largest numbers. It is believed that the solution to this problem is either "Graham's Number" or 6.[3]

A Scale Is a Scale Is a Scale

Let's say you weigh yourself twice, once on a spring scale (like the one in your bathroom) and once on a balance scale (the kind where you put the object to be weighed on one side and an equal weight on the other). Both scales say you weigh 150 pounds. Now go to the equator and do the same thing. Then to the North Pole. Then down in a deep mine. Then in an airplane.

The balance scale will always tell you that you weigh 150 pounds (i.e., your weight equals that of the "150 pound weight"). The spring scale, however, will tell you that you weigh more than this at the equator and less at the other places.

Here's why. The balance scale measures your mass, and this is the same, relative to the weight on the other side of the balance, at all places. The spring scale measures your weight, the pull of the earth on you (or, more accurately, the pull of you and the earth on each other; but the former is much greater, unless you have *really* been overdoing it at the dinner table).

Because the earth is slightly fatter at the equator than at the poles (as a result of "centrifugal force"), there is more of the earth pulling on you at the equator and less at the other places. When you are deep in the mine, for example, the portion of the earth above you is pulling in the other direction from the portion of the earth that is below you, reducing the pull downward.

Out of Nothing

About 13.8 billion years ago, a hole in an infinite eternal nothingness gave birth to a singularity, a possibly infinitely small universe that contained (in the form of energy) all of the energy and mass now contained in the universe it has become. Out of nothingness we can make a hole through which can be born all logic and all mathematics. From nothingness we can make the set containing nothing (i.e., the empty set), which we write "{ }" or "0" and call "zero." We can then make the set "{ { } }." We can call this set "1." We can then make the set containing the set containing the set containing nothingness. We can call this 2 . . . .

Is it the same nothingness from the nothingness in which the universe was born? Did mathematics (including the extensive mathematics that seem to have nothing to do with the physical world) somehow stream out of the same hole as the universe? The obvious answer is no: one is a physical nothingness of a virtual vacuum, while the other is an intellectual construct. But, somehow, the obvious answer does not seem obviously correct. And we do know that all of nature follows the rules of mathematics and is explicable in mathematics, even if not all of mathematics has physical counterparts. Perhaps such mathematics is a blueprint for what the physical universe will become.

Or maybe mathematics is relevant to the physical universe only in the ways we already know about. Whatever the case, such things are fun to think about.

The Powers That Be

You may have wondered why a number raised to the zero power is one. You may have thought that a number raised to the zero power should be itself (because the number is not multiplied by anything, so it should just remain itself). But then you probably then remembered that a number raised to the *first* power is itself, and you would not want a number raised to the power 0 to equal the same number as the number raised to the power 1.

But why couldn't we have a number raised to the power 0 be 0, just as a number multiplied by 0 is 0? It is true that this would not be terribly satisfying (because the number is *not* being multiplied by zero, but by itself zero times). But, as W. V. Quine points out, the strongest reason why we cannot permit a number raised to the 0 power to be 0 is this:

> We want n^{m+1} always to be n^m times n (for these are the same thing). That is, $2^{3+1} = 2^3 \times 2$.
>
> If we take m as 0, we get
>
> $n^1 = n^0$ times n, which is the same as
>
> $n = n^0$ times n.
>
> So n^0 must be 1.

The First Uninteresting Number

According to David Wells's *The Penguin Dictionary of Curious and Interesting Numbers,* 39 is the first uninteresting (natural) number—which, of course, makes it interesting.

Hey, This Is Easy

Hey, this is easy. 1, 2, 4, umm.

> Wells also points out that $n^n + 1$ is equal to a prime number when n is 1, 2, or 4.
>
> $1^1 + 1 = 2$
>
> $2^2 + 1 = 5$
>
> $4^4 + 1 = 257$

If there is another prime that is generated by this method, it has, at minimum, three hundred thousand digits.

Judy Is a Bank Teller, But Is She a Feminist?

Judy is thirty-three, unmarried, and quite assertive. A magna cum laude graduate, she majored in political science in college and was deeply involved in campus social affairs, especially in antidiscrimination and antinuclear issues. Which statement is more probable?

(a) Judy works as a bank teller.
(b) Judy is active in the feminist movement and works as a bank teller.

Most people, correctly perceiving that Judy may well be a feminist, choose (b). But note that if (b) is true, then (a) must also be true. If Judy is a bank teller and a feminist, Judy must be a bank teller. There is, however, a possibility that (a) is true and (b) is not. That is, Judy is a bank teller and not a feminist. Thus, (a) is more probable.

This is a wonderful example, created by psychologists Amos Tversky and Daniel Kahneman, of the way in which people come to incorrect conclusions because they substitute inappropriate mental models for logical thinking.[4]

A Quotation Whose Source I Have Lost

Here is a quotation whose source I have long since lost:

> All that exists or could have existed or could come to exist—in the mind or in potential or in reality—is the set of all sets, which has the same structure as the set of all complex one-dimensional subspaces of a complex infinite-dimensional Hilbert space.

I have only the foggiest idea what this means, but it sure *sounds* nifty.

Notes

1. For simplicity's sake, this is the way Goldbach's conjecture is almost invariably presented. Apostolos Doxiadis, in his very fine novel *Uncle Petros and Goldbach's Conjecture* has a footnote giving a more accurate description:

> In fact, Christian Goldbach's letter (to Euler) in 1742 contains the conjecture that "every integer can be expressed as the sum of three primes." However, as (if this is true) one of the three such primes expressing even numbers will be 2 (the addition of three odd primes would be of necessity odd, and 2 is the only even prime number), it is an obvious corollary that every even number is the sum of two primes. Ironically, it was not

Goldbach but Euler who phrased the conjecture that bears the other's name—a little known fact, even among mathematicians.

2. We also know that every odd number is the sum of not more than 300,000 primes because we need merely add 1 (here considered a prime) to the primes that sum to an even number to get the primes that sum to next odd number.
3. David G. Wells, *The Penguin Dictionary of Curious and Interesting Numbers*, 1998.
4. From John Allen Paulos, *Once upon a Number: The Hidden Mathematical Logic of Stories*.

5

Believe It or Not

Start Counting

However large the number of physical things (remember, the number of particles in the universe is way, way below 10^{100}), the numbers appearing in analyses of the universe and its parts is much, much greater.

In refuting the Austrian physicist and philosopher Ludwig Boltzman's claim that molecules in an enclosed space will forever increase, mathematician Henri Poincaré showed that the molecules will return to their original state, thereby decreasing entropy. For a gas of just forty liters, this will happen in a brief $10^{\text{trillion trillion}}$ seconds. (The universe is only 10^{17} seconds old.[1])

Long Time Comin'

In his delightful book *The Secret Lives of Numbers,* mathematician George G. Szpiro reminds us of the distinction between convergent and divergent series. Convergent series approach a limit. For example: 1 + 1/2 + 1/4 + 1/8 . . . approaches 2.

Divergent series, on the other hand, go wandering into infinity. However, they may do so very slowly. For example, the "harmonic series" 1 + 1/2 + 1/3 + 1/4 . . . takes 178 million steps before it reaches 20.

A Nonsucker's Bet

You have a half-dollar piece and a silver dollar. I have a quarter. We toss all three coins. Each tail counts zero. Each head counts its value in points. The player with the greater number of points in each round wins the value of the *other* player's coin(s).

Thus, if I toss a head with my quarter and you toss a head with your half-dollar and a head with your silver dollar, you win 25 cents (my quarter).

I would be a fool to play. Right?

Wrong. The results will balance out.

Each time you win, you win only 25 cents. Each time I win, I win $1.50. You will win six times as often, but I will win six times as much each time I win.

ME	YOU	RESULT
H	H H	You win 25 cents.
H	H T	You win 25 cents.
H	T H	You win 25 cents.
T	T H	You win 25 cents.
T	H H	You win 25 cents.
T	H T	You win 25 cents.
T	T T	No one wins.
H	T T	I win $1.50.

As long as the coins are adjacent on the doubling sequence ($25 \times 2 = 50$; $50 \times 2 = 100$), no coin is duplicated, and one player has at least one coin and the other at least two, the sides will always be even.

So What?

Consider this: Primes of the form $3n + 2$ are more numerous than primes of the form $3n + 1$ until you get to 608,981,813,029, at which point primes of the form $3n + 1$ take the lead. The lead switches back and forth, but for an infinity of values, $3n + 1$ wins.

I Had Only Gotten Up to the One-Inch 3-Cube

John Derbyshire, author and novelist, makes a fascinating point. Did you know you can put the Empire State Building into a 1-inch cube? You just need enough dimensions.

Longest corner-to-corner diagonal in a one-inch 2-cube (i.e., square): square root of 2

Longest corner-to-corner diagonal in a one-inch 3-cube (i.e., cube): square root of 3

Longest corner-to-corner diagonal in a one-inch 4-cube (i.e., "hyper-cube"): square root of 4

Longest corner-to-corner diagonal in a one-inch 5-cube: square root of 5

And so on.

So the long diagonal of a one-inch n -cube is sqr(n). If n is about 200 million, the one-dimensional version of the Empire State Building will slide in nicely!

Perfect Numbers

A "perfect number" is a number for which the integers that divide evenly into the number (other than the number itself) also add up to the number. The smallest perfect number is 6. Six is evenly divisible by 1, 2, and 3, and $6 = 1 + 2 + 3$. In addition, $6 = 1 \times 2 \times 3$, making it what might be termed a "perfect, perfect number." (I bet that it can be easily proven that 6 is the only "perfect, perfect number, but not so easily that I can do it.) The next perfect number is 28 ($28 = 1 + 2 + 4 + 7 + 14$). It is, obviously, not a "perfect, perfect number," as $1 \times 2 \times 4 \times 7 \times 14 = 784$, not 28.

There are only about forty perfect numbers. The largest has over 130,000 digits. All are even numbers. Is there an odd perfect number? No one has discovered one or proven that there could not be one.

Incidentally, a "sublime" number is one that has the number of divisors (that leave no remainder) that is perfect and the sum of these divisors is perfect. There are only two known sublime numbers. One is 698, 655,567,023,837,898,670,371,734,243,169,822,657,830,773,351,885,970, 528,324,860,512,791,691,264. The other is 12 ($1 + 2 + 3 + 4 + 6 + 12$, the six divisors, sum to 28). The first two perfect numbers are 6 and 28.

Pretty, Very Pretty

As Euler proved, the only solution in positive integers for $a^b = b^a$ is $2^4 = 4^2$.

Probability of Hitting a 1-in-*X* Event in *X* Tries?

Most people know that the probability of tossing a coin and having it come up heads is 1 in 2 (0.50); of rolling a four with a six-sided die is 1 in 6 (0.166+); and of landing on an eight on a 1–10 wheel is 1 in 10 (0.1). However, when you ask these people the probability of rolling at least one four in six rolls or landing at least once on eight in ten spins, they will usually answer 1 in 2 (or 50 percent or fifty-fifty or "even money"—these are all the same thing).

As you can see below, when the possibilities are four or more, the probability of at least once attaining the 1-in-n event in n attempts is always roughly 2 in 3 (.66+), or, more accurately, a bit above 63 percent.

(In other words, the probability of rolling at least one four in six rolls of the die (a one-in-six chance) or having the spinner land on "8" in ten spins (a one-in-ten chance) is a bit below two-thirds. But even if the wheel has a million numbers and you win only if the spinner lands on number 562,194, if you spin the spinner a million times, there is about a 63% chance that you will land on 562,194 at least once.

If there are only two or three possibilities, then the likelihood is somewhat higher. Take a coin as an example. The likelihood of tossing a head is, of course, one in two. If you toss the coin twice, the possibilities are

head-head
head-tail
tail-head
tail-tail

As you can see, the chance of tossing at least one head in two tosses is 3 in 4 (0.75); only tail-tail fails to get a head. Once we get to a 1-in-6 case, like the die, we fall below the 2-in-3 likelihood and soon approach the 63+ percent likelihood, but never go below it.

The table below shows the 1-in-*n* probability of at least one occurrence of an event with a probability of 1 in *n* in *n* events: for example, selecting the Jack of Hearts in fifty-two attempts, with the deck being reshuffled each time.

1 in 2 (coin)	0.75
1 in 3	0.70+
1 in 4	0.68+
1 in 5	0.67+
1 in 6 (die)	0.66+
1 in 10 (1–10 wheel)	0.65+
1 in 12	0.65–
. . .	
1 in *n*	0.63+

With every increase in *n*, the probability decreases, but it decreases by a smaller amount each time. The limit is $1 - 1/e$, which is >0.63, but <0.64. (The *e* is our old friend from compound interest.) Thus, if you try an event a million times that has a one-in-a-million probability, the probability of hitting the event at least once is >0.63 and <0.64.

No matter how large n is, there will always be a better than 0.63 probability of hitting a 1-in-n event in n tries.

You will have a 0.50-0.50 chance of hitting n in just about $0.7n$ tries. Thus, if a wheel is numbered 1–100 and you want the spinner to land on number 64, you have a 0.50-0.50 chance of landing on number 64 at least once if you spin the wheel seventy times and a 0.63+ chance if you spin it a hundred times.

Imagine How Good He Was When He Wasn't Dying

This is perhaps the personal anecdote one most often hears in informal mathematical circles. I have got to admit that it never struck me as being as spellbinding or hilarious as it does others, but here it is for the sake of completeness.

When the great mathematician Srinivasa Ramanujan was on his deathbed, he was visited by G. H. Hardy. The number theory expert told Ramanujan that he had hoped that the number of the taxi would be an interesting one that would cheer up Ramanujan. Alas, he had to inform Ramanujan that the number was an uninteresting 1,729.

"Oh, no," cried Ramanujan, as he leaped from his deathbed, "that is a very interesting number; it is the smallest number that is expressible as the sum of two cubes in two different ways" ($1^3 + 12^3 = 1 + 1{,}728 = 1{,}729$ and $9^3 + 10^3 = 729 + 1{,}000 = 1{,}729$).

Ramanujan was a scientist of numbers. His fascination with their nature led him to care more about discoveries about numbers than proof that a discovery was, in fact, a discovery of a "true fact" of numbers. His astonishing intuition and intelligence did indeed lead to many discoveries of the highest order.

But It Is Less Than $10^{70{,}000{,}000{,}000{,}000}$ Plus 1

The largest number I have ever seen in a discussion of the physical world is $10^{70{,}000{,}000{,}000{,}000}$, which John Barrow (*The Constants of Nature: The Numbers That Encode the Deepest Secrets of the Universe*) estimates as the number of possible wiring connections between the neurons of the brain. This is, of course, the number of *possibilities*, not a number referring to any actual physical reality, and it dwarfs the number of possible combinations of DNA ($10^{3{,}480{,}000{,}000}$).

The Medical Test

Consider this: You are given a medical test for a dreaded disease—say, Floogle's Sliding Nose Disease—and you get a positive result. This test is 98 percent accurate for both positives and negatives. Do you kiss your nose "good-bye"?

Do not do it. You are more likely to lose your nose doing that than from Floogle's Disease. Here's why.

Say 10,000 people take the test for Floogle's, and 50 of them actually have the disease. Of the 50 who have Floogle's, 49 will correctly get positive results (i.e., 98 percent of 50 = 49), and 1 will incorrectly get a positive result (i.e., 2 percent of 50 = 1).

Of the 9,950 who do not have Floogle's, 9,751 will correctly get negative results (i.e., 98 percent of 9,950 = 9,750), and 199 will incorrectly get positive results, false-positive test results (i.e., 2 percent of 9,950 = 199). (These numbers have been rounded off very slightly.)

Thus, of the 248 people who get positive test results (49 + 199), only 49 actually have Floogle's disease. About 80 percent of those people who got positive results *do not* have Floogle's disease.

This does not mean that the test is of little use. With the assumptions we have made (i.e., that the test is 98 percent accurate, for both positives and negatives, and 50 of 10,000 have Floogle's), a positive result tells us little, but a negative result makes it very, very likely that one does not have the disease.

Haircut Anyone?

Bertrand Russell, philosopher, historian, mathematician, logician, created a logical paradox that struck at the heart of all logic.

In the town of Whatsgoingonhere (not the name that Russell used), the barber shaves all those and only those who do not shave themselves. Who shaves the barber?

If the barber shaves himself, he breaks the rule that he shaves only those who do not shave themselves.

If the barber does not shave himself, he must, according to the rules, shave himself.

Something (Actually, Everything) out of Nothing

Where do integers come from? Well, there are a lot of different answers to this question, but here is an especially elegant one.

A number is a set. For example, 4 is the set containing four things. Whether these things are chairs, beliefs, planets, or left-handed bowlers under three feet tall is irrelevant. It is the "fourness" that is the set that is the number 4.

Now, an empty set is merely a set with nothing in it—no chairs, no beliefs, no planets, no left-handed bowlers under three feet tall. Call this set "zero" and represent it as { }.

Where sets of 4 differ in their being four chairs or four beliefs or four planets or four left-handed bowlers, there is only one empty set; whether it is empty of chairs or beliefs or planets or left-handed bowlers is irrelevant. It is empty.

Now consider a set that contains the empty set. Represent it as {{ }}. Call this set "1."

The set that contains the set that contains the empty set—{{{ }}}—we will call "2."

And this goes on, forever. Out of nothing (the empty set), we (well, actually, the great mathematician John von Neumann) create all the integers.

Simple, Simply Tricky

You have two coins, and you are going to toss each once. What is the probability that you will toss matching coins (i.e., two heads or two tails)?

It is simple. Note that there are four equally probable possibilities:

Head Head
Head Tail
Tail Head
Tail Tail

Thus, there is a 50 percent chance that you will throw matching coins.

The following is not so simple; it is simply tricky. You have four loose socks in a bag, two blue and two red. You pull out two socks. What are the chances that you will pull out two matching socks?

The problem seems identical to the coin toss. There are four possibilities:

Blue Blue
Blue Red
Red Blue
Red Red

Two of the four possibilities give two socks of the same color. So this is the same problem as that of the coins and has the same 50 percent probability, right? Well, actually, no.

In the coin problem, the events were independent (i.e., unconnected). Tossing, say, a head on the first toss does not affect the probability of the second toss coming up heads or coming up tails. The possibilities for the second toss remain at one head or one tail.

However, in the sock problem, your selecting a blue sock changes the possibilities for the second pull. It is no longer two blues and two reds, but one blue and two reds because you have used up one of the blues. In the coin problem, tossing a head did not use up a head.

Incidentally, the coin problem is analogous to roulette, in which each spin of the wheel is independent (so that there can be no system that wins at roulette). The sock problem is analogous to blackjack, in which cards are used up, making a system possible, but very, very difficult.

It Looks Random

Random numbers play a crucial role in science. Like "time," "cause," and other basic conceptions of the constituents of reality, the meaning of "random" seems too obvious to require thought—until we think about it.

When we think of random numbers, we think of numbers that possess no pattern, sequences for which we can do no better than guessing when we predict the next number in the sequence. So, for example, it must be true of such a number that each digit tends to appear about one-tenth of the time (in no predictable order), each pair of digits about one-hundredth of the time (if this is not the case, then there is a patterned bias built in). A number with this quality is called "normal." (Note that no rational fraction possesses this quality because such fractions always repeat: 1/3 is 0.3333 . . ., and 1/97 seems random for 96 digits and then keeps repeating those 96 digits. Also, the constraints we have mentioned must take into account the number of digits in the string; a string of ten digits is clearly not likely to be random if it is 2222222222. However, in a string of a quadrillion digits, 222222222 will appear many times.)

However, there is another quality that we tend to automatically associate with randomness: not merely is there no pattern, but there is no easy algorithm. Thus, while pi has no pattern (at least that we can discern in pi's first billion digits), the sequence is generated by an easy rule (an algorithm for pi), and this enables us to predict every digit.

The two forms of "randomness" are in conflict. As Martin Gardner suggests, write down ten one-digit numbers. If you wrote down any number more than once, the sequence fails to meet the first criterion (i.e., it is biased toward the number you wrote down more than once). However, if you write each number just once, you satisfy the first criterion, but fail to meet the second (because, given nine digits, one can predict the tenth). This sort of conflict is inherent in all attempts at randomness, and there is no such thing as a sequence that meets both criteria.

There is, in other words, a limit precluding our attaining that which we tend to assume is true of what we mean by "random."

Pack Those Bins

Say you have a bunch of packages of various sizes and shapes, and you have a number of identical bins to put them in. Surprisingly, there is no algorithm known that guarantees that you attain the most efficient packing (i.e., the minimum number of bins). This is known as an NP-complete problem, and, as in the case of the traveling salesman problem (see page 2), it is probable that there is no *efficient* algorithm (because the number of bins grows arithmetically, but the time to solve the problem grows geometrically). To get the ideal packing, you pretty much have to try all the possibilities, and if there are more than a few bins, this would take all the computers in the world a billion, trillion, or more years.

However, if you simply fill the first bin as much as you can (paying no attention to package size), then do the same with the next bin, and so on, you will need only about 70 percent more bins than if you tried every possibility. If you pack in order of decreasing size (put the largest remaining package in the bin, then the next largest that will fit, etc.), you will come within 20 percent of the minimum number of bins.

Note

1. Paul Davies, *About Time: Einstein's Unfinished Revolution.*

6

Science, Math . . . and Baseball

How the World Series Were Won (through 1998)

There are thirty-five ways of winning a World Series (i.e., there are thirty-five possible combinations of wins and losses in a World Series).[1] Seven of these—two six-game combinations and five seven-game combinations—have never occurred.

For example, the seven-game combination in which the winning team wins games one, two, six, and seven and loses games three, four, and five has occurred twice (Minnesota's victories in 1987 and 1991). Note, incidentally, that the one, two, six, and seven combination we use as an example is the only one in which a team lost *any* three consecutive games and won the series (i.e., no team has won a series by winning games one, two, three, and seven (and losing four, five, and six); by winning one, five, six, and seven (and losing two, three, and four); or by winning four, five, six, and seven (and losing one, two, and three).

It is interesting to compare the percentages of 4-, 5-, 6-, and 7-game World Series with what would be expected in a "coin-toss series."

A coin-toss series is one played by two teams identically matched in every way, who play on a neutral field, and are unaffected by previous games; in other words, neither team has any advantage. The two teams are a "heads" team and a "tails" team, with the former winning if the coin lands heads and the latter winning if the coin lands tails. The series is won by the team that first wins four tosses.

One might suspect that, because in a real World Series one team is often better than the other, the series would tend to be shorter than the coin-toss series with their equally matched teams.

The cynic, on the other hand, might expect the players in the real World Series to stretch the series to increase profits. (Players only share in the receipts for the first four games, but it does not take that much cynicism to wonder whether the increased owners' profits from a longer series might not indirectly profit the players.)

No. Games/ No. of Combinations	Games Won/ (Games Lost)	Years Combination Occurred	Total Games
4 Games/ 1 Combination	1, 2, 3, 4	'07, '14, '22, '27, '28, '32, '38, '39, '50, '54, '63, '66, '76, '89, '90, '98, '99	17
			No. of four-game series = 17
5 Games/ 4 Combinations	1, 2, 3, 5, (4)	'10, '37, '70	3
	1, 2, 4, 5, (3)	'08, '16, '19*, '29, '33, '88	6
	1, 3, 4, 5, (2)	'05, '13, '41, '43, '49, '61, '74, '84	8
	2, 3, 4, 5, (1)	'15, '42, '69, '83	4
			No. of five-game series = 21
6 Games/ 10 Combinations	1, 2, 3, 6, (4, 5)	Never	0
	1, 2, 4, 6, (3, 5)	'95	1
	1, 2, 5, 6, (3, 4)	'17, '30, '53, '80	4
	1, 3, 4, 6, (2, 5)	'18, '77, '93	3
	1, 3, 5, 6, (2, 4)	'06	1
	1, 4, 5, 6, (2, 3)	'20**	1
	2, 3, 4, 6, (1, 5)	'11, '35, '36, '48, '59, '92	6
	2, 3, 5, 6, (1, 4)	Never	0
	2, 4, 5, 6, (1, 3)	'23, '44, '51	3
	3, 4, 5, 6, (1, 2)	'78, '81, '96	3
			No. of six-game series = 22
7 Games/ 20 Combinations	1, 2, 3, 7, (4, 5, 6)	Never	0
	1, 2, 4, 7, (3, 5, 6)	'72	1

No. Games/ No. of Combinations	Games Won/ (Games Lost)	Years Combination Occurred	Total Games
	1, 2, 5, 7, (3, 4, 6)	'47	1
	1, 2, 6, 7, (3, 4, 5)	'87, '91	2
	1, 3, 4, 7, (2, 5, 6)	'12***, '67	2
	1, 3, 5, 7, (2, 4, 6)	'09, '62, '97	3
	1, 3, 6, 7, (2, 4, 5)	'34, '73	2
	1, 4, 5, 7, (2, 3, 6)	'60, '64	2
	1, 4, 6, 7, (2, 3, 5)	Never	0
	1, 5, 6, 7, (2, 3, 4)	Never	0
	2, 3, 4, 7, (1, 5, 6)	Never	0
	2, 3, 5, 7, (1, 4, 6)	'31, '75	2
	2, 3, 6, 7, (1, 4, 5)	'26, '82	2
	2, 4, 5, 7, (1, 3, 6)	'45, '57	2
	2, 4, 6, 7, (1, 3, 5)	'24, '40, '46, '52	4
	2, 5, 6, 7, (1, 3, 4)	'03(**), '25, '68, '79	4
	3, 4, 5, 7, (1, 2, 6)	'55, '56, '65, '71	4
	3, 4, 6, 7, (1, 2, 5)	'21(**), '86	2
	3, 5, 6, 7, (1, 2, 4)	'58, '85	2
	4, 5, 6, 7, (1, 2, 3)	Never	0
			No. of seven-game series = 35

*These three series were best five of nine that, had they been a best-four-of-seven series, would have been won in the win-loss combination noted. All three combinations have also occurred in regular best-four-of-seven series.

**In 1920, Cleveland won a best-five-of-nine series by winning games 1, 4, 5, 6, and 9. Had this been a best-four-of-seven series, Cleveland would have won by winning games 1, 4, 5, and 6, and so this is considered here to have been the equivalent of a seven-game series in which the winning team won games 1, 4, 5, and 6. This combination has not occurred in any other World Series.

***This series included a tie game (called on account of darkness). If the tie game is ignored, the win-loss combination is that noted. This combination has occurred also in a best-four-of-seven series that had no ties.

1903: First Series1904 & 1994: No SeriesTotal # of Series: 95

To compare the real World Series to a coin-toss series one must, of course, calculate the expected lengths of coin-toss series. To do this, one does *not*, as would seem reasonable, base the calculation of probabilities on the thirty-five real possibilities listed in the table above on pages 52 through 53. The most common error in probability assessment is forgetting that a calculation of probabilities must be based on equally likely possibilities. The thirty-five real possibilities are *not* equally likely.

* * *

There are 128 equally likely sequences of wins and losses (heads and tails) in a seven-game series. Ninety-three of these cannot happen in a real World Series (e.g., W, W, W, W, L, L, L), but they are counted when figuring probabilities. (The 128 is 2^7, or $2 \times 2 \times 2 \times 2 \times 2 \times 2 \times 2$; each 2 represents a game in which a team can win or lose.)

Thus, the expected probabilities in the coin-toss series (rounded to nearest full percentage point) are

13%—about one in eight—of the coin-toss series will be a four-game series.
25%—about one in four—of the coin-toss series will be a five-game series.
31%—about one in three—of the coin-toss series will be a six-game series.
31%—about one in three—of the coin-toss series will be a seven-game series.

How does this compare with the real World Series (rounded to nearest full percentage point)?

18% (17 of 95; about one in six) of real World Series have been a four-game series.
22% (21 of 95; about one in four-and-a-half) of real World Series have been a five-game series.
23% (22 of 95; about one-in-four-and-a-half) of real World Series have been a six-game series.
37% (35 of 95; about four-in-ten) of real World Series have been a seven-game series.

The only really significant divergence from randomness is the fewer six-game series and greater number of seven-game series than would be expected if skill, home advantage, and so on, played no role. This is probably owing primarily to the human factor: the one who is threatened with being eaten fights harder than the one anticipating the meal.

There is, however, an important additional reason: one team has a 3–2 advantage in home games in the first five games, and this is evened up in the sixth game. The team that is home in the sixth game is more often the team that is down, rather than up, 2–3, and home teams win more often than they lose. Therefore, the probability of a series going to the seventh game if it reaches the sixth is greater than the 0.50–0.50 of the coin toss. (Home teams down 2–3 have won sixth games nearly twice as often as they have lost them because the present configuration of home and away games was introduced.)

Science and Mathematics

It is an axiom of contemporary science that there is no a priori knowledge of the world.

Some aptitudes may require no empirical experience (i.e., human beings clearly have the ability to distinguish one thing from another; if everything looked the same to an infant, there would be no possibility of human development.) However, discovery of facts about the world always require at least an initial empirical observation (from daily life or formal experiment). (That is, the aptitude to make distinctions does not become knowledge that there are distinctions in the real world until the infant observes the real world.)

In other words, science cannot be deductive in the way mathematics can. Mathematics defines its own world. Thus, if A is a real number greater than B and B is a real number greater than C, A is greater than C. There is no need of empirical observation because A is defined as being greater than B, and B is defined as being greater than C. Note that what would rule this out as science is not our deducing logically that A is greater than C, but the absence of even an initial observed empirical fact.

This point is important because people occasionally—when distinguishing the (deductive) logic of mathematics from the inductive logic of science—understate the importance of deductive logic's ability to discover new facts. In reality, it is very often the case that we learn new facts through deduction from known—empirically discovered—old facts. (This is because at least one initial fact *must* be known from empirical observation that this is science, not mathematics.)

Let us take an example from everyday life, assuming that we all know that the World Series is won by the first team to win four games and that the first, second, and, if they are necessary, sixth and seventh games are played in one team's ballpark, while the third, fourth, and if it is necessary, fifth game are played in the other team's ballpark.

Note that the above information does not represent discovered facts but our construction of a World Series, something approaching a definition more closely than a discovered fact. Which team is home in which game is more a rule than a fact. Note also that we can deduce from our construction that a series must go the full seven games for a team to lose any three consecutive games and win the series and that it must go the full seven games for the home team to win every game.

Now, following the 1987 World Series, the newspapers reported what we will call

> Fact One: This was the first World Series in which every game was won by the home team.

To know this fact the newspaper men had to look to the reported observation of empirical reality (i.e., the record book); there is no purely logical way to know that no team had previously won the World Series by winning its four home games.

"So what?" you may well be asking at this point. The "so what" is this: when going to the record book, an astute reporter could have noticed

> Fact Two: No team had ever lost *any* three consecutive games and won the World Series.

Fact two is superior to fact one in that (a) it is more general and (b) it permits one to *logically deduce* fact one. It is more general because it tells us not merely that before 1987 no team had won games 1, 2, 6, and 7 (the only way a team can win by winning its home games), but also that no team ever won a World Series by winning games 1, 2, 3, and 7 or 1, 5, 6, and 7 or 4, 5, 6, and 7. Thus, for example, fact two tells us that no team has ever won a World Series by winning all of its away games (which also requires winning games 1, 2, 6, and 7).

Fact two tells us more about reality than fact one while also telling us fact one.

Fact one does not permit us to logically deduce fact two. The fact that 1987 was the first time that the home team won every game does not tell us that it had never been the case that the away team won every game or that no team had won the World Series by winning games 1, 2, 3, and 7 or 1, 5, 6, and 7 or 4, 5, 6, and 7.

We're not talking $e = mc^2$ here, but we are talking about the same relationship of logic and science. We live in the same physical universe that Einstein did, and we can learn about that physical universe only through the same logic and science that Einstein did.

A Game of Dominoes

British author Simon Singh, who specializes in writing about mathematics and science, in his excellent *Fermat's Enigma* gives a wonderful example of the difference between the deductive method of mathematics and the inductive method of science.

Picture a regular chess board or checkerboard with two opposing corner squares cut off the two opposing white squares. This leaves 32 white squares and 30 black squares. Now take 32 dominoes, each the size and shape of two squares. (Ignore the numbers on the dominoes; they are irrelevant.)

Here is the question: can you place the 32 dominoes on the board in such a way that the board is exactly covered? (A domino must cover two squares exactly, so it cannot be cut in two or placed on two diagonally touching squares.)

The scientific method of doing this would be to test a good number of combinations and, finding that none of them precisely covered the board, *tentatively* conclude that it is impossible to do so. But this would never give certainty because someone might subsequently find a combination that does the trick.

The mathematical method is this:

> Note that the two removed opposing corners were the same color (as they had to be; whether the two were both white or both black does not matter).
>
> Note that no two (nondiagonally) adjacent squares are the same color.
>
> Each domino must cover two *adjacent* squares.
>
> Therefore, the first 30 dominoes cover 30 white squares and 30 black squares, leaving 2 black squares.
>
> Thus, you are always left with 2 squares of the same color.
>
> But the final domino must, like all the others, go on squares of two different colors (because it must go on *adjacent* squares, and adjacent squares are always two different colors)
>
> Therefore, the 32 dominoes can never exactly cover the board.

Hardly Ever Happens

It is not uncommon for mathematicians to find that a guess (a conjecture) or an educated guess (a hypothesis) turns out to be incorrect. It even occasionally happens that a proof whose conclusion is true is found to be faulty (i.e., not prove what it claims to be true.) In such cases, a later proof virtually always proves the conclusion.

What almost never happens is that a seeming proof turns out to be faulty and its conclusion turns out to be false. Marcus du Sautoy, professor of science and mathematics at Oxford, gives an example that comes close. The French mathematician Augustin-Louis Cauchy believed he had a proof that all geometric solids satisfied a certain equation. It turned out that Cauchy's equation failed for a cube with a hole in its center.

I know what you are thinking: So Cauchy did have a proof for solids without holes, and if a solid has a hole it is not a solid. No mathematician buys this, and, anyway, a proof that covered solids with holes was soon discovered.

How We Measure Our Ignorance

While there are various views of precisely what probability is, and while probability on a quantum level is a property of nature herself, in practical terms, probability is usually a measure of ignorance. Consider, for example, your chance of guessing the suit of the bottom card in a deck. It is one in four.

But now consider the situation if you get a peek at the bottom card—just enough of a peek to see that it is black. You now know that the suit must be a spade or a club, so your chance of guessing the suit is one in two. The only thing that has changed is your ignorance, which has been cut in half.

Maybe if I Used a Hammer

Remember those 15 Puzzles you played as a kid? They were small, black-and-white square things with fifteen small squares numbered one to fifteen (in consecutive order) and one empty space. You could slide the numbered squares around to make different combinations of numbers.

You may remember that the companies that made the puzzles offered tremendous prizes to anyone who could move the pieces around and end up with certain combinations. You never could manage to get these combinations. But it was not your fault. They were impossible.

You would have saved a lot of time if you had known the simple way of finding out whether a combination was possible.

Count the number of times a number is followed by a lower number until a higher number is reached. So, for example, the desired combination is 15-14-13-12-11-10-9-8-7-6-5-4-3-2-1. This combination is possible because there is an even number (14) of decreases. Now consider 1-15-2-14-3-13-4-12-5-11-6-10-7-9-8. This is impossible because there is an odd number (7) of decreases.

A Map Is Something They Used to Give You at a Gas Station

It had been suspected for centuries that no map on a flat surface required more than four colors for every district (i.e., country, state, etc.) to have a different color from any bordering country. We refer here not just to a map of the real world, but a map of any possible configurations of countries (as long as no country or state is—like Michigan—divided into two or more separate parts and no countries meet at only a single point).

However, as of 1976, no one had been able to prove that there could not be a configuration requiring five colors. Then, in 1976, Wolfgang Haken and Kenneth Appel, colleagues and mathematicians at the University of Illinois, discovered a proof—a proof that required twelve hundred hours of computer time. This proof required some brilliant mathematics in addition to the computer time, but words cannot describe how much mathematicians despise it. The proof cannot be checked by a mathematician in a hundred lifetimes (except by rerunning the programs). In other words, the four-color proof is a bit lacking in the snappy elegance of Euclid's proof of an infinity of primes.

You Try That Bridge, and I Will Try This One

A river divided the city of Konigsberg. In the river were two islands. Island A had a bridge to Island B and a bridge to each bank. Island B had the bridge to Island A and two bridges to each bank.

For many years the residents of Konigsberg added interest to their Sunday strolls by attempting to cross every bridge once and only once. (It was not necessary to end the stroll at the starting point.) No one accomplished the task, but some residents kept trying.

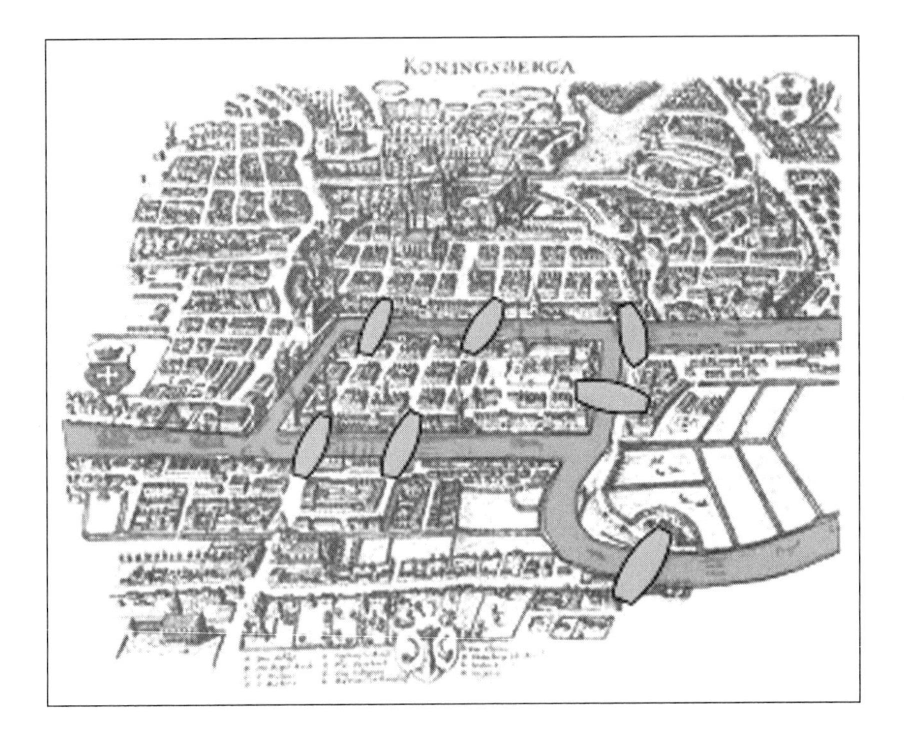

Had they consulted the great mathematician Leonhard Euler they could have saved a lot of shoe leather. Consider that the three bridges of Island A could be crossed in four ways:

1. Cross all three bridges by walking *toward* Island A.
2. Cross all three bridges by walking *away from* Island A.
3. Cross two bridges by walking *toward* Island A and one bridge by walking *away from* Island A.
4. Cross two bridges by walking *away from* Island A and one bridge by walking *toward* Island A.

Obviously 1 and 2 are impossible: you cannot walk *to* a place three times without walking *away from* it, and you cannot walk *away from* a place three times without walking *to* it.

Choices 3 and 4 are each possible, but because the number of bridges of Island A is odd (i.e., three), the stroll must either begin (4) or end (3) at Island A.

Now consider one of the two banks of the river. It also has three bridges, so the stroll must either begin (4) or end (3) on this bank.

Fine. The stroll might begin on Island A and end on the bank or begin on the bank and end on Island A.

But wait. The other bank *also* has three bridges. The stroll must begin or end here too. Because the stroll cannot begin at two different places or end at two different places, the stroll is impossible. If one more bridge were appropriately added (which, in fact it has been), the stroll would be possible.

In general, whatever the number of "points" (in this case islands and banks) and paths or "lines" (in this case, bridges), you can cover all paths without retracing your steps *only* if the number of points (in this case, islands and banks) at which an odd number of paths meet is 0 or 2.

Not to Bother

Since ancient times, many people have attempted to solve the following three problems using the tools of Euclidian geometry, invoking only straight lines and circles or parts of circles and using only a straightedge and a compass:

1. Doubling of the cube: create a cube with sides double that of a given cube.
2. Trisection of an angle: divide an angle into three equal angles.
3. Square the circle: create a square with the same area as a given circle.

You would be better off spending your time training a cow to jump over the moon. We do not *know* that there could never be a cow that can jump over the moon. We *do* know that the above problems cannot be solved. There are proofs that they are unsolvable.

You Want Logical, I'll Give You Logical

Okay, okay. You want *one* standard logic puzzle. Here is a classic. About 2 or 3 percent get this one right within ten minutes. It has no "catch" and is particularly elegant in that *every* fact given is necessary.

On a train, Smith, Robinson, and Jones are the fireman, brakeman, and engineer, but Smith is not *necessarily* the fireman, Robinson is not *necessarily* the brakeman, and Jones is not *necessarily* the engineer. Also on the train are three businessmen: Mr. Smith, Mr. Robinson, and Mr. Jones. We know these facts:

1. Mr. Robinson lives in Detroit.
2. The brakeman lives exactly halfway between Detroit and Chicago.
3. Mr. Jones earns exactly $20,000 dollars a year.
4. The brakeman's nearest neighbor, one of the passengers, earns exactly three times as much as the brakeman.
5. Smith beats the fireman at billiards.
6. The passenger whose name is the same as the brakeman's lives in Chicago.

What are the names of the fireman, the brakeman, and the engineer?

While we will not consider this a brainteaser (it takes too long), the answer is given at the end of the book.

The Twelve Days of Christmas

Over "The Twelve Days of Christmas," "my true love gave to me" 364 gifts. (On the first day, my true love gave to me: 12 + 11 + 10 + 9 + 8 + 7 + 6 + 5 + 4 + 3 + 2 + 1 = 78. On the second day: 11 + 10 + 9 + 8 + 7 + 6 + 5 + 4 + 3 + 2 + 1 + 66, etc. 79 + 66 + . . . = 364.)

Here is a simpler way:

Twelfth day: 12 + 11 + 10 . . . 1	= 78
Eleventh day: 78 minus the twelfth day's 12	= 66
Tenth day: 66 minus the eleventh day's 11	= 55
Ninth day: 55 minus the tenth day's 10	= 45
. . .	
First day: 3 minus second day's 2	= 1

The 3*N* + 1 Problem

This problem has two rules:

1. When you have an odd number, triple it and add 1.
2. When you have an even number, halve it.

Now, select any positive whole number and proceed.

Say you select 6. Because 6 is even, halve it—giving 3. Because 3 is odd, triple it and add 1—giving 10. Because 10 is even, halve it—giving 5. Because 5 is odd, triple it and add 1—giving 16. Because 16 is even, halve it—giving 8. Because 8 is even, halve it—giving 4. Because 4 is even, halve it—giving 2. Because 2 is even, halve it—giving 1. Because 1 is odd, triple it and add 1—giving 4. Because 4 is even, halve it—giving 2. Because 2 is even, halve it—giving 1. Because 1 is odd, triple it and add 1—giving 4.

Notice anything? The sequence has entered an unending 4-2-1-4-2-1 . . . loop. "No big deal," you say. How about this? The same is true for every positive whole number through 27,000,000,000,000,000. Is it true for every positive whole number over 1,000,000,000,000? No one has been able to prove it is or to find an exception. But it is not for a lack of trying.

As George G. Szpiro points out in his fascinating book *The Secret Life of Numbers*, if there is an exception, it must be a number greater than 27 quadrillion (i.e., 27 with fifteen zeros) and must have at least 275,000 steps before settling down to 1.

How Does 0.99999 . . . Equal 1?

You may have been a bit dubious about the claim that 0.99999 . . . is the same as 1. But consider this: 0.33333 . . . plus 0.33333 . . . plus 0.33333 . . . equals 1/3 plus 1/3 plus 1/3, which equals 1.

The Traveling Salesman Problem

You might think that there would be no practical application of mathematics more easily accomplished than this: a traveling salesman must visit thirty-three cities, and he needs to know the shortest route that gets him to every city at least once.

You saw way back on page 2 that not only can the mathematician not give you the shortest route, he cannot even tell you whether there is any way other than trial and error for finding the shortest route with certainty. And when there are more than a few cities, even a fastest computer using only trial and error would make for a hopeless task; fifty cities, for example, would have to check 10^{62}, a task that would take millions of years. There *are* algorithms guaranteeing a route not more than a few percent longer than the shortest route.

This problem, the "packing-bin" problem discussed in Chapter 5, and a million other problems all resist the discovery of a practical algorithm. If an algorithm for even *one* of these problems were found, it would work for all of them. If it can be shown that there can be no *efficient* algorithm (see "Pack Those Bins," on page 49) for even one of these problems, then there can be no such algorithm for any of them.

Can You Do This?

This simple equation has never been solved in positive integers, nor has anyone ever proved that there is a solution or that there is no solution (other than 0):$(x + y + z)^3 = xyz$

Spelling Integers

You have to count to one thousand before the letter "a" is required to spell an integer.

Taking a Random Walk

You see an endless line of squares of concrete:

−3	−2	−1	+1	+2	+3

You toss a coin. Whenever it is heads, you go to the right. Whenever it is tails, you go to the left.

Intuition tells you that on any long walk it is likely that you will be to the left of the center line about as often as to the right; in other words, there will be a lot of "switching sides" (i.e., crossing the center line).

But this is not the case. No matter how long the walk, the most likely situation is that you never switch sides. The second-most likely is that you switch sides only once. The third-most likely is that you switch sides twice, and so on.[3]

The Most Common Error

The sum of two even numbers is an even number. The sum of two odd numbers is an even number. The sum of an even number and an odd number is an odd number. Thus, if you select two numbers at random, the odds are 2–1 that the sum will be even. Correct? No, incorrect.

The most common error in probability, an error that can very subtle, is that which considers possibilities as equally probable when they are not. When you select two numbers at random, there are *four*, not three, equal possibilities (many people incorrectly conflate 2 and 3):

1. even-even = even sum
2. even-odd = odd sum
3. odd-even = odd sum
4. odd-odd = even sum

Thus, it is fifty-fifty; you will get as many odd sums as even.

If you do not believe this, get a random-numbers table. Add the first two digits in each of the first hundred numbers. In the overwhelming majority of the hundred cases, it will be obvious that percentages of even and odd sums are converging on fifty-fifty. However, in a very small percentage of cases, one actually does get even sums approaching or even surpassing two-thirds. This is expected. Those who get this result are urged to sum another two hundred pairs. In nearly every case, these additional sums will make it clear that even and odd sums are converging on fifty-fifty. In such extraordinarily rare cases that it would not be worth mentioning except that we know as surely as we know anything that they will arise in a statistically predictable (tiny) percentage of the cases, there will still be two-thirds of the sums coming up even. If this should happen, do another four hundred numbers. If there is *still* no convergence on half and half, we will know that the messiah has come (or returned, depending on your point of view) and you are him.

Cubing Cubes

The number of cubes needed to sum to 239 is nine (i.e., $239 = 4^3 + 4^3 + 3^3 + 3^3 + 3^3 + 3^3 + 1^3 + 1^3 + 1^3$). No higher number takes more than eight cubes, and it has been conjectured (but not proved) that there is some number after which it never takes more than seven cubes.

Mathematical Literacy Test

A recent test of the mathematical literacy of high school students included a question that was both elegant and interesting.

Let us say that a person is not taxed on the first $10,000 of income, but is taxed 6 percent on all income above $10,000. How much would one have to make for the tax to be 6 percent of the entire income?

The answer is that no matter how large the income over $10,000, if it is taxed at 6 percent, the omitted $10,000 will always make the tax rate on the total income less than 6 percent.

Notes

1. The formula for determining that there are thirty-five possibilities is $CrN = N!/R!(N - R)!$ (i.e., # combinations = $(7 \times 6 \times 5 \times 4 \times 3 \times 2 \times 1)/(4 \times 3 \times 2 \times 1)(3 \times 2) = 35$

   The formula does not, of course, tell you what the thirty-five sequences are; that must be done by rote.
2. Martin Gardner, *Mathematical Circus.*

7

The Case for Logic

Intuition versus Logic

William Dunham, in his excellent book *The Mathematical Universe: An Alphabetical Journey through the Great Proofs, Problems, and Personalities,* gives a beautiful and simple example of a situation in which intuition and common sense are incorrect and logic is correct.

Consider a swimming pool that has a perimeter of 240 feet, say a swimming pool that is 100 feet long and 20 feet wide. Intuition tells us that the shape does not affect the area as long as the length of the perimeter remains unchanged (i.e., you can change the shape from a nonsquare rectangle to a square or even a hexagon without changing the area). The logic of mathematics demonstrates that this is incorrect.

Our first swimming pool is 100 feet by 20 feet. Its perimeter is 240 feet (100 + 20 + 100 + 20), and its area is 2,000 square feet (100 × 20).

Now let's change the swimming pool to a square shape with the same perimeter of 240 feet. The square is 60 feet on each side. The swimming pool now has an area of 3,600 feet (60 × 60). So the length of the perimeter is the same, but the area is different.

It is even possible, as Dunham points out, for a longer perimeter to contain less area than a shorter one. For example, a swimming pool 2 yards wide by 12 yards long has a perimeter of 28 yards, but an area of only 24 square yards. A square pool measuring 5 yards on each side has a shorter perimeter of 20 yards, but a larger area of 25 square yards.

Pi-Eyed

Pi (π), the ratio of the circumference of a circle to the circle's diameter, is an unending number that begins 3.14159 . . . . For practical use, the number must be rounded off at some point.

A century ago, as students are often told, a legislator from Indiana attempted to have pi declared to equal precisely 3.2. Ever since, this has served as the quintessential example of an absurd attempt to impose a human wish on mathematical and scientific truth, the equivalent of a law declaring the earth to be flat.

While I would never admit it to a mathematician, this legislative foray—while no doubt dopey to the nth degree and potentially making *a lot* of creaky bridges and leaning buildings—never struck me as sufficiently stupid to carry the moral weight the story is meant to carry. The story is told as if the legislator were denying a truth where the practitioner of *applied* mathematics was asserting it. (In pure mathematics pi need not be rounded off.) This *would* be the case had the legislator declared pi to be a rational number, rather than the transcendental number that it is. *Then* he would have deserved the degree of ridicule the story attempts to convey; he would have committed a mathematical felony. (Do not worry about what a transcendental number is; all that matters here is that it is not a rational number.)

But the legislator did not deny truth where the mathematician asserts it; they both deny truth, the legislator only quantitatively more so—a misdemeanor. Pi is not 3 nor 3.14159; it is a number beginning with 3, 3.14, 3.141, 3.1415, 3.14159, and so on.

Small World

Many Americans have had the experience of traveling in a far-off American city and meeting someone who is a friend of a friend of theirs. As astonishing as this invariably *seems*, it is to be expected. Itheil de Sola Pool of MIT found that the probability is over 60 percent that any two randomly selected Americans can be linked by two or fewer intermediaries.

This finding is based on the assumption that the average American *personally* knows a thousand people well enough to recognize them and call them by name: friends from the sixth grade, all the doctors you have gone to more than once, and so on. (Celebrities known only through the media do not count.) Thus, if I randomly choose an American adult (the president of the United States or an alcoholic on a wharf in Southern California or a taxidermist in Idaho) and you are an American, the probability is better than 6 in 10 that you know someone who knows someone who knows the person I randomly selected. These figures apply to the continental United States as a whole and impose no further constraints.

If we are more rigorous in our definition of "know," so that the average person knows only a hundred people, the number of intermediaries increases by two or three. If, on the other hand, a further constraint is added—say, both you and your friend are college graduates—the likelihood that you know someone who knows someone who knows your college graduate friend is greatly increased.

In practice, this last is the case when, for example, you realize that you know someone known to the person next to you on the plane. You and your travel acquaintance are both wealthy enough to fly, possibly residents of the same city, and so forth. What is surprising is not that you both know the same person, but that you discovered this fact during a short plane ride.

If all this seems too astonishing to believe, think about how quickly the network tree of acquaintances grows. You know a thousand people by name and face, and the thousand people know an average of a thousand each. That is a million. Those million know an average of a thousand each. That is a (US) billion. Even with duplications removed, that is *a lot* of people (four times the American population) to whom you are this closely "connected." It does not seem so surprising that these people include 60 percent of the American population.

To see this all more clearly, picture a clockface that, instead of the 12 numbers around the face, has 150 million numbers. (It is a *really* big clock face, and the numbers are *really* small.) Each number is "connected" to ("knows") 100 other numbers, with no 2 numbers sharing any other numbers. Because each number is connected to 100 other numbers, any 2 numbers will be connected to 10,000 numbers (100 × 100). Each of these 10,000 numbers is connected to 100 other numbers, so just three degrees of separation makes a million (100 × 100 × 100) connections. Add another 100 connections and we are up to 100 million (100 × 100 × 100 × 100). With just one more 100 numbers we have 10 billion connections. (There are only about 6 billion people in the world.)

However, there is a very big *however*. In our clockface example, we assumed that all of (the numbers equivalent to) Alan's friends were different from (the numbers equivalent to) Bob's friends, and all of (the numbers equivalent to) Bob's friends were different from (the numbers equivalent to) Charlie's friends, and so forth. In real life, many of Alan's friends are also friends of friends of Alan's (i.e., Bob is Charlie's friend); thus, there are many, many duplications, which add nothing to the connectedness.

Nonetheless, as Cornell University mathematicians Duncan Watts and Steven Strogatz have shown, a very few long-range connections (say, the one friend you have in India, the one patient an Australian doctor had who now lives in England, etc.) reduce the number of connections needed to connect any two people more dramatically than would seem possible, and this generates the reality Itheil de Sola Pool found.

The small-world situation is still far from settled. There are statisticians who doubt the randomness of the selected samples (as well as one or another of the assumptions made), with one going so far as to suggest that the entire matter is an urban legend. Future research will likely settle the issue.

Gravity

Gravity, at least in our universe, follows what is known as the inverse square law. Say you and a rock are floating in space x miles away from each other. When you double the distance, you reduce the gravitational force to $1/x^2$. This is because we live in a universe of three spatial dimensions.

If we lived in a universe of four spatial dimensions, things would be quite different. As the British cosmologist and astrophysicist Martin Rees points out, the reduced gravitational force would be $1/x^3$, and the planets would be slowed down sufficiently to cause them to plummet into the sun. Good-bye to us. (Whether there could be a universe of four spatial dimensions is not known.)

Rees explains why the inverse law holds. Think of your distance from the rock as the radius of a sphere, with the rock at the center (or a sphere with you at the center). As you increase the radius of the sphere, you spread the lines of gravitational force over a larger area, thereby diluting and weakening them as described above.

A Categorization of Number Systems

A number system that always permits addition, multiplication, and subtraction without introducing a new kind of number is called a "ring." The positive and negative integers constitute a ring. A system that permits these *and* division is called a "field." The integers do not constitute a field because one cannot always divide an integer by an integer and get an integer (i.e., 29 divided by 3 is not an integer). The rational numbers—numbers expressible as a fraction of two integers, such as 1/2 or 743/627 *do* constitute a field because dividing a rational number by a rational number always gives a rational number.

Nonplussed

Julian Havil is a retired mathematics master of England's Winchester College. In his *Nonplussed! Mathematical Proof of Implausible Ideas*, Havil gives two interesting examples of nonintuitive truths:

1. Consider two equally talented tennis players, each of whom wins 90 percent of the games in which he serves.

   In tennis, scoring goes 15, 30, 40, win, with the winner having to be two points ahead. This is what is counterintuitive: a server who has a 40–30 lead is less likely to win the game than a server in a game that is just beginning (i.e., 0–0), and this is true for purely mathematical reasons.

   This is why: on any point, the nonserver has only a 10 percent chance of winning. The probability that he will win three points in a row (to win the game he is losing 40–30) is greater than his winning any game.
2. If Andy is taller than Bob and Bob is taller than Charlie, then Andy *must* be taller than Charlie. This is pure logic.

   You might think that this is true for any term substituted for "taller than." But consider the kids' game of Rock, Paper, Scissors. Scissors cuts (i.e., beats) paper, paper covers (i.e., beats) rock, and rock breaks (i.e., beats) scissors (where Charlie is *not* taller than Andy).

   Where "taller than" is linear, Rock, Paper, Scissors is circular.

And a One, and a Two

Counting one number per second,

it will take	to count to
1 second	1
17 minutes	1,000 (thousand)
12 days	1,000,000 (million)
32 years	1,000,000,000 (billion)
32,000 years	1,000,000,000,000 (trillion)
32 million years	1,000,000,000,000,000 (quadrillion)
32 billion years	1,000,000,000,000,000,000 (quintillion)

Thirty-two billion (US terminology) years is about two or more times the current age of the universe, depending on the cosmological theory one uses to estimate the age of the universe.

Hey, Wait a Minute

$a = b$
So, $ab = a^2 = b^2$
So, $ab + (a^2 - 2ab) = a^2 + (a^2 - 2ab)$
So, $a^2 - ab = 2a^2 - 2ab$
So, $1(a^2 - ab) = 2(a^2 - ab)$
So, $1 = 2$

What is going on here? Substitute numbers for a and b and you will find that the next-to-last step is $1 \times 0 = 2 \times 0$ or $0 = 0$, not $1 = 2$. In other words, to get to $1 = 2$, both sides of $1(a^2 - ab) = 2(a^2 - ab)$ were divided by 0. That is not allowed.

What Is So Rare as 10^{10923}?

In everyday life, we usually assume that there are few of something that is "rare." People who can run a mile in under three minutes and fifty seconds are not merely rare in frequency, but in absolute numbers. This assumption is valid because we live in a finite world.

However, it is worth remembering that in mathematics "rare" usually means "exceedingly low frequency." Rare numbers may be infinite in number. For example, numbers evenly divisible by 10^{10923} are very rare in frequency. But there are an infinite number of them. There are, of course, numbers that are both relatively and absolutely rare. For example, 2 is the only even prime, and 6 is the only perfect, perfect number. (See "Perfect Numbers," on page 43.)

The Holy Grail

If God promised to whisper one secret into your ear, it is a good bet that most mathematicians would ask for the simple algorithm for the primes (i.e., a *simple* formula that would identify all the primes and only the primes, a formula that is both necessary and sufficient for the identification of primes.) For it is likely that such an algorithm would take us to the heart of the heart of numbers.

There *is* such an algorithm if we drop the word "simple." If you take a number, find the factorial of one less than that number, add one to the result, divide by the number, and find that the remainder is 0, the number is prime. Otherwise it is not prime. For example, start with 5; $4 \times 3 \times 2 \times 1 = 24$; $24 + 1 = 25$; $25/5 = 1$ with no remainder. So 5 is prime. The problem is that it takes virtually as long to do this than to test for primality by trial and error.

As mathematician Tobias Dantzig points out in *Number*, while the above is called the Wilson Index, it was Leibnitz who proved that the condition is necessary, and Lagrange who proved, a hundred years later, that it is sufficient.

Incidentally, you may have been wondering why 1 is not considered a prime. (It is considered a unique number, neither prime nor composite.) The reason is this: A crucial mathematical theorem, the Fundamental Theorem of Arithmetic, states that every positive integer (1 excepted) is the product of one, and only one, set of primes or 1 and a prime. For example, $11 = 1 \times 11$ and $36 = 2 \times 2 \times 3 \times 3$. If 1 were considered a prime, then, for example, 11 would equal 1×11 or $1 \times 1 \times 11$ or $1 \times 1 \times 1 \times 11$, and so on.

The Nature of a Rational Number

You will remember that a rational number is one that can be expressed as a fraction (i.e., a "ratio"). All integers are rational because they can be expressed as themselves over 1 (e.g., $7 = 7/1$). In some cases, it is easy to tell that a number is not rational. The square root of any integer that is not a whole number is not rational (i.e., the square root is not rational.) The square roots of 2, 3, 5, 6, 7, and 8 are all nonrational. The square root of 9 is rational because it is the whole number 3.

8

A Short Math Miscellany and a Final Thought . . .

Mathematical Truth versus Scientific Truth

Thousands of years ago, mathematicians wondered whether there is a greatest prime number after which no other number is prime. As mentioned previously, it turns out that there is no largest prime. No matter how large a prime you specify, there is always a larger prime. Later, we shall see the stunningly elegant and simple proof of this.

In Chapter 2 we introduced the idea that in mathematics, "simple" does not mean "easy to do" but rather "relatively few steps from question to answer." A simple mathematical proof may go undiscovered for millennia and be discoverable only by the greatest of geniuses. This proof did not go through all the possibilities, which is obviously impossible. Nor did it try a lot of possibilities and figure, "Hey, if we haven't found one by now, there must be none." Such inductive reasoning, the method that has made *science* so successful, is not permitted in mathematics. Mathematics requires proof that something *is* true, or *could not* be true, of numbers unimaginably large. There is no "very probably" in mathematics.

For a bit of the feel of what a proof is, consider this question: can two consecutive whole numbers both be even? Even without working through the proof, you know that the answer is no. If one number is even, the next will be odd. You do not have to try the impossible task of checking all the numbers. You do not even have to check out every number from 1 to 100. You simply have to point out that when a number is even, the next number will, when divided by 2, leave a remainder of 1. A number divided by 2 that leaves a remainder of 1 is not even. Therefore, there cannot be two consecutive whole numbers that are both even.

You might be thinking, "Yeah, but 'even' is practically *defined* as meaning two consecutive integers cannot both be even." That is true, but, as we shall see, it is true of *all* mathematics. The difference is not that it is true of some proofs and not others. It must be true for proof to be possible. The difference between two proofs is that one is obvious and the other is *only* obvious to a great mathematician.

A mathematical truth differs from a scientific truth in another crucial way. A scientific truth—a fact about the real world—is always tentative. If Mount Everest floats away tomorrow for seemingly no reason, we would not say that, because such an event is seen as impossible by our theories of gravity, Mount Everest could not have *really* floated away. We would go back to the drawing board and attempt to develop a theory that succeeds in every way the old theory did *and* also explains why Mount Everest floated away. Likewise, while we are pretty sure that we will never discover a Brazilian tribe of people with two heads, our confidence is based only on probability (an *extremely* high probability).

A Pair of Squares

Between two randomly chosen consecutive squares, say, four and nine, no one has ever failed to find at least one prime (in this case, five and seven). Are there any consecutive squares that fit within the gap between consecutive primes (i.e., that have no prime in their span)? No one has found such a pair of consecutive squares or proved that there could not be one.

Similarly, between any integer and its double (other than one) there is a prime. This *was* proved by math genius Paul Erdös at age twenty.

* * *

Is there a solution to this equation (no zeros permitted)?

$$(x + y + z)^3 = xyz$$

You guessed it. No one knows.

The above two entries are based on *Tomorrow's Math: Unsolved Problems for the Amateur* by C. Stanley Ogilvy.

Get to Work

I guess by now you are ready to tackle a problem no one has solved, and you have half an hour to kill before *Law and Order*. So try this.

A transcendental number is a type of number whose nature was discovered after the types of numbers mentioned in the introduction. It is a number, like *e*, that is not a solution to an algebraic equation with integer coefficients.

Most numbers—though not those you and I are familiar with—are transcendental. Deciding whether a specific number is transcendental is difficult beyond belief.

An Incandescently Beautiful Equation

Posted on *August 4, 2011* by *Deskarati*

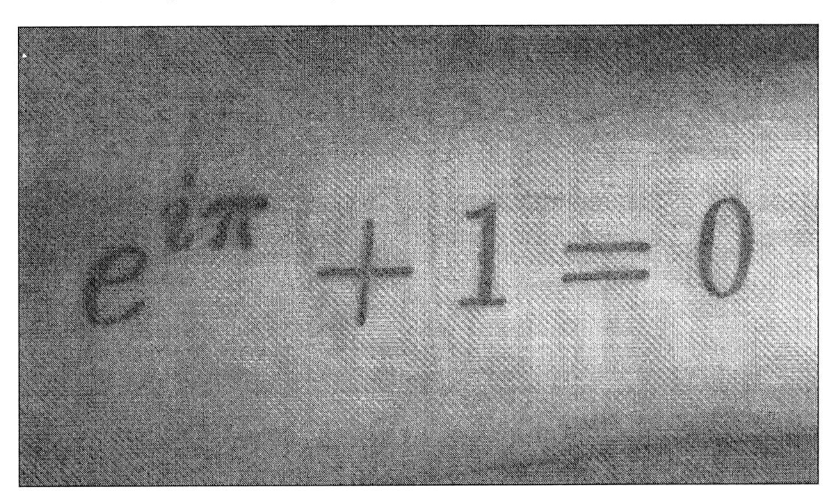

This formula by *Leonhard Euler,* known as Euler's identity, was called "the most remarkable formula in mathematics" by theoretical physicist *Richard Feynman,* for its single uses of the notions of addition, multiplication, exponentiation, and equality, and the single uses of the important constants 0, 1, *e, i* and π. In 1988, readers of the *Mathematical Intelligencer* voted it "the Most Beautiful Mathematical Formula Ever". In total, Euler was responsible for three of the top five formulae in that poll.

The Vision of an Exceptional Mind

Pi is, among other things, the ratio of the circumference of a circle to the circle's diameter.

The *e* is, among other things, the base of the natural logarithms.

The basic "imaginary" number *i* is the square root of –1.

The 0 is the godhead of all numbers.

Euler showed that the relationship among these is:

$$e^{i\pi} + 1 = 0$$

Many a mathematician claims to have seen the face of God in this equation.

But Why?

The sum of the sequence of odd numbers always equals the square of the number of numbers.

1 (1 number) = 1sq
1 + 3 (2 numbers) = 2sq
1 + 3 + 5 (3 numbers) = 3sq
1 + 3 + 5 + 7 (4 numbers) = 4sq
1 + 3 + 5 + 7 + 9 (5 numbers) = 5sq
1 + 3 + 5 + 7 + 9 + 11 (6 numbers) = 6sq
1 + 3 + 5 + 7 + 9 + 11 + 13 (7 numbers) = 7sq
1 + 3 + 5 + 7 + 9 + 11 + 13 + 15 (8 numbers) = 8sq

Why?

Consider the middle number of each sequence. In the case of 1 + 3 + 5 + 7 + 9 + 11 + 13, the middle number is 7. In the case of a sequence with an even number of numbers, the middle number is the average of the two middle numbers; that is, the middle number of 1 + 3 + 5 + 7 + 9 + 11 is 5 plus 7, all divided by 2, or 6.

Note that in each sequence the number of numbers times the middle number equals the sum of the numbers. For example, in 1 + 3 + 5 + 7 + 9 + 11 + 13 the middle number (7) times the number of numbers (7) equals 49, or 7^2. The next sequence is 1 + 3 + 5 + 7 + 9 + 11 + 13 + 15. Here the number of numbers (8) times the middle number (8) equals 64. Every increase to the next odd number works in the same way.

But Why Is a Minus Times a Minus a Plus?

It is obvious that a positive times a positive is a positive; negatives never enter the picture. And it is easy enough to see why a positive times a negative is negative (or vice versa); ten times minus (i.e., a debt of) a hundred dollars is minus a thousand dollars, and a hundred times minus ten dollars is minus a thousand dollars.

Nearly everyone knows that a negative times a negative is positive, but most people take this on authority and never quite understand why. In his book *The Art of Mathematics*, Jerry P. King gives a beautiful explanation.

Consider the number x. Assume that we have proven that there exists an x that can be defined as follows and so define it:

$x = ab + (-a)(b) + (-a)(-b)$ (definition of x)
So: $x = ab + (-a)[(b) + (-b)]$
$= ab + (-a)(0)$
$= ab + 0$
$= ab$

And:

$x = ab + (-a)(b) + (-a)(-b)$ (definition of x)
So: $x = [a + (-a)]\mathrm{b} + (-a)(\mathrm{-b})$
$= (0)(b) + (-a)(-b)$
$= 0 + (-a)(-b)$
$= (-a)(-b)$

Thus: $ab = (-a)(-b)$

(That is, each solution equals x, so solutions equal each other.)

And: $x = (-a)(-b)$
Therefore: $ab = (-a)(-b)$
Which is the same as: $(-a)(-b) = ab$
Or: (in other words) a minus times a minus is a plus.

Getting up to Speed

In 1993, the IBM Research Center in Yorktown Heights, New York ran 566 processors—each effectively a computer—for a year in a test of the theory of quantum chromodynamics, the heart of our understanding of particle interaction. The computers performed over a hundred million billion calculations. The theory passed the test.

Less Is More

In Roman times, the subtractive method of writing numbers was rarely used; 4 was IIII, not IV, and 9 was VIIII, not IX. The subtractive method was not common until the late Middle Ages.

Both methods can be seen on a present-day clock that uses Roman numerals: 4 is usually written IIII, but 9 is never written VIIII.

A Final Thought

Euclid

Old Euclid drew a circle
On a sand-beach long ago.
He bounded and enclosed it
With angles thus and so.
His set of solemn greybeards
Nodded and argued much
Of arc and of circumference,
Diameters and such.
A silent child stood by them
From morning until noon
Because they drew such charming
Round pictures of the moon.

–Vachel Lindsay

Twenty-Five of the World's Greatest Brainteasers, Plus Two

(Answers to Nonbrainteasers Follow the Answers to the Brainteasers)

Brainteasers

#1: A Wimbledon-type (elimination) tennis tournament has 256 first-round entrants. How many total matches will there be in the tournament? You have three seconds to answer.

#2: An acquaintance tells you that he has two children. You ask him if at least one of the children is a boy and he says yes. What is the probability that he has two sons? (This is a mathematical, not a scientific, question, so assume that equal numbers of boys and girls are born and live to adulthood.)

#3: You have a scale that gives an accurate reading to the ounce. There are seven piles, each containing seven weights. The seven weights in any given pile are equal to each other. The first pile has weights labeled "one pound," the second pile has weights labeled "two pounds," and so on. One of the piles contains weights that are each either one ounce too light or one ounce too heavy. You may weigh whichever weights you like. How many weighings are required to determine which weights are mislabeled and whether they are an ounce heavy or an ounce light?

#4: I have heard it said that the earth is smoother than a billiard ball (i.e., if you expanded the billiard ball to the size of the earth, the highs and lows of its surface would be greater than the mountains and valleys of the earth). I have no idea whether this is true, but for this problem let us assume that the earth itself is perfectly smooth. Let us also assume that the circumference of the earth is exactly 25,000 miles (i.e., 132 million feet).

A piece of string 132 million feet long is stretched taut around the earth. You add a foot of string to the string, so the string is now 132,000,001 feet long. If you prop up the string equidistantly around the earth, how tall will the props have to be? Not only will you find this answer surprising, you will also find another surprise given with the answer.

#5: How many ways are there to change a dollar?

#6: You are on the TV game show, *Let's Make a Deal*. There are three closed doors: behind one, a car has been randomly placed; behind the others are goats. You win whatever is behind the door that you end up choosing, and you want to win the car. The host, Monte Hall, plays fair; he does nothing to trick you or to help you.

Monte says, "You choose a door. Then I'll open a different door, one that I know has a goat behind it. Then you may stick with your original choice or may switch to the other closed door."

Should you: (A) stick with your original choice? (B) Switch? (C) It makes no difference?

#7: This is a relatively easy one to boost your spirits. In a two-mile race, going twice around a one-mile oval track, your car averages thirty miles an hour for the first lap. How many miles per hour must you average on the second lap to average sixty miles an hour for the entire race?

#8: How many randomly selected people would you have to invite to a dinner party to be certain that

At least three all know each other, or[1]

At least three all do not know each other.[2]

#9: An anthropologist and spouse go to a party at which there are ten people, including the anthropologist and spouse. Everyone at the party, except possibly the anthropologist, shakes hands with a different number of people (i.e., one person shakes hands with no one, one with one person, one with two people, etc.).

Who is the anthropologist married to (i.e., how many people did the anthropologist's spouse shake hands with)?

#10: A hat contains three cards: one white on both sides; one white on one side and red on the other; and one red on both sides. You pick a card and place it on the table with white showing. You do not look at the other side.

I size up the situation and offer to give you 3–2 odds ($1.50 to $1.00) that the other side is white. Should you take the bet?

#11: Three men pay ten dollars each for a thirty-dollar hotel room. The concierge later realizes that he should have charged only twenty-five dollars for the room and returns five dollars to the men, who then tip him two dollars. So, each man paid nine dollars (the original ten minus one returned), and the concierge got two dollars. Where did the other dollar go?

#12: This may be the oldest of all brainteasers, and you may well know it in one of its many forms. But it is a beauty and worth including for those who are reading it here for the first time.

There are two doors. One is the "door of life." The other is the "door of death." In front of each door is a man. One of the men—you do not know which—always tells the truth. The other man always lies.

Your life hangs on the one question you are permitted to ask one of the men. After asking the question, you choose a door. If you choose the door of life, you live (and, what the hell, marry the princess, win billions of dollars, and get to meet Michael Jordan). If you choose the door of death, you are a memory.

What question do you ask?

#13: This is one of Martin Gardner's favorites of the easier brainteasers.[3]

Miss Green, Miss Black, and Miss Blue are out for a stroll. One is wearing a green dress, one a black dress, and one a blue dress.

"Isn't it odd," says Miss Blue, "that our dresses match our last names, but not one of us is wearing a dress that matches her own name?"

"So what?" replies the lady in black.

What was the color of each lady's dress?

#14: This brainteaser is at least 150 years old.

Your boss offers you a choice of two bonuses: (1) fifty dollars after six months and a semiannual increase of five dollars, or (2) one hundred dollars after a year and an annual increase of twenty dollars. Which bonus will prove more lucrative?

#15: This is a great one discovered by Martin Gardner—one that shows how God forgot to wire our brains so that we know probability by intuition.

You are playing roulette. You pick any triplet you like (i.e., black-black-black, red-red-black, etc.). I then pick a different triplet. The game ends when someone wins, but we will play as many times as you like.

I will give you three dollars each time you win; you give me two dollars each time I win. There are even odds, and I am giving you 3–2. It is a great bet for you, right?

You choose black-black-black, and I choose red-black-black. You will win if the first three spins are black. Should you take the bet?

#16: Here is an old chestnut that should be easy, but nearly everyone gets it wrong, even after they have heard the answer ten times:

If a hundred chickens eat a hundred bushels in a hundred days, how many bushels will ten chickens eat in ten days?

#17: This famous brainteaser, "Dudeney's Cigar Puzzle," rivals the tennis brainteaser for elegance and requires absolutely no mathematical knowledge.

You and a friend are sitting at a normal-sized, square (topped) table. You each have more cigars than are needed for this problem. The cigars, which are identical, are normal cigars (i.e., normal size, flat at one end and pointed at the other). You place a cigar on the table. Your friend then places a cigar anywhere on the table he likes, as long as it does not touch your cigar. You then place a cigar anywhere on the table you wish, as long as it does not touch any other cigar. And so it goes. The winner is the last person to successfully place a cigar.

Who wins, and why?

Hint: The "why" is crucial. There are acceptable arguments for either person's winning.

#18: You have two barrels. Barrel 1 is filled with 10 gallons of red paint. Barrel 2 is filled with 10 gallons of blue paint. You take a gallon of red paint from barrel 1 and pour it into barrel 2, mixing the red and blue paint thoroughly. You then take a gallon of the mixture in barrel 2 and pour it into barrel 1.

Which is now greater: the red paint in barrel 1 or the blue paint in barrel 2?

#19: Two poles, each fifty-feet tall, are constructed somewhere on a football field. A sixty-foot rope is strung from the top of one to the top of the other. The lowest point of the rope is twenty feet from the ground. How far apart are the poles?

#20: This is a great one from Martin Gardner (who heard it from University of Oregon emeritus professor of psychology Ray Hyman, who, in turn, read it in a monograph by Gestalt psychologist Karl Duncker).

One morning, exactly at sunrise, a Buddhist monk began to climb a tall mountain. The narrow path, no more than a foot or two wide, spiraled around the mountain to a glittering temple at the summit.

The monk ascended the path toward the temple at varying rates of speed, stopping many times along the way to rest and to eat the dried fruit he carried with him. He reached the temple shortly before sunset. After several days of fasting and meditation, he began his journey back along the same path, starting at sunrise and again walking at variable speeds with many pauses along the way. His average speed descending was, of course, greater than his average climbing speed.

Is there any way that there could be a spot along the path that the monk will occupy on both trips at precisely the same time of day. If not, prove that there cannot be such a spot.

#21: This is a tough one. You have two pieces of string (different lengths and different thickness). Each burns at varying rates, and the two pieces burn at different varying rates. (For example, when you light the end of the first piece, it might burn a foot in the first minute, half a foot in the second minute, two feet in the third, whatever. When you light the end of the second piece of string, it might burn at entirely different rates.)

All that the two strings have in common is that they each burn completely in an hour. How can you use these strings to measure forty-five minutes?

#22: You and two other people, Tom and Harry, who are both very quick but not quite as quick as you, sit facing each other. Each of you has a red or a white hat on. Each person can see the hats of the other two people, but not his own. You see that Tom and Harry have red hats. Can you deduce the color of your hat?

#23: You come to a fork in the road. One path leads east, the other west. One of the paths leads to heaven, the other to hell, but you do not know which is which.

At the fork are two men; one always tells the truth, and one always lies. You do not know which man is which.

You can ask one question to learn which path is the path to heaven. What do you ask?

#24: This is a nifty one from Marilyn vos Savant.

It is taking you too long to get to your workplace. You figure that you will leave later and arrive earlier. How do you accomplish this?

#25: You have a girlfriend who lives downtown and a girlfriend who lives uptown. The uptown and downtown trains alternate, first one and then the other. So, you figure, you will leave to chance which girlfriend you will visit each night, taking whichever train comes in first at whatever time you happen to get to the subway station. This way you will visit each girlfriend the same number of times.

But it does not work out that way. About five times out of every six, the uptown train comes. How come?

Answers to the Brainteasers

Answer #1: If you were permitted more time, you could add the matches played in the first round, the second round, etc.: 128 + 64 + 32 + 16 + 8 + 4 + 2 + 1. This is the way in which the intelligent nonmathematician would go about solving the problem. But solving the problem this way does not capture the elegance of the mathematical insight.

What is this mathematical solution? Consider this: There is one and only one loser per match and one and only one match per loser. Every player except the tournament winner (who does not lose) loses once and only once. Therefore, the number of matches equals the number of players who lose. There are 255 players who lose, so there are 255 matches in the tournament. (For the dubious, 128 + 64 + 32 + 16 + 8 + 4 + 2 + 1 = 255.)

If you did not have this insight, do not worry. Neither do many mathematicians.

Answer #2: You know that this is not the obvious answer, 1 in 2 (i.e., fifty-fifty). This would be the answer if the acquaintance stated that the *first* child (or the *second* child) was a boy. If the acquaintance had said that the first was a boy, he would have left only two possibilities:

Boy-Boy
Boy-Girl

However, when the acquaintance answers that *at least one* is a boy (without telling you whether it is the first or second child), he slips in some new information. The new information is that there are *three* possibilities:

Boy-Boy
Boy-Girl
Girl-Boy

In two of these cases, the other child is a girl. In only one is the other child a boy. Therefore, the probability that the other child is a boy is 2 in 3, not 1 in 2.

If you still do not believe this, take a random-numbers table and consider only the first two digits of each number. Consider odd digits as boys and even digits as girls. Circle all the numbers in which one or both digits are odd (i.e., "at least one is a boy"). Finally, for all of *these*, compare the number of numbers in which one digit is even (i.e., a boy and a girl) with the number of numbers in which both digits are odd (i.e., two boys). You will find approximately twice as many of the latter.

This brainteaser is sometimes stated: "A couple has two children. One is a boy. What is the probability that the other is a boy?" Unfortunately, this form, which is undeniably more natural and felicitous, leaves an ambiguity concerning the method of randomization and the possibility that 1 in 2 can be considered a correct answer. The form given here avoids this problem.

Answer #3: One. Put on the scale: one one-pound weight, two two-pound weights, three three-pound weights, four four-pound weights, five five-pound weights, six six-pound weights, and seven seven-pound weights. If the scale reads one ounce over any exact pound weight, the one-pound weights are heavy. If it reads x pounds and two ounces, the two-pound weights are heavy, and so on. If the scale reads x minus one pound and fifteen ounces, the one-pound weights are light. If it reads x minus one pound and fourteen ounces, the two-pound weights are light, and so on.

Answer #4: Given the enormous length of the string, you probably assumed that the props would be tiny, say, one-millionth of an inch. However, the props must, in fact, be an impressive 1.91 . . . inches tall.

Here's why. The circumference of a circle equals pi (3.14 . . .) times the diameter of the circle. This is true for every circle everywhere.[4] If you add 12 inches (a foot) to the circumference of a circle, you add 12/3.14 . . . (i.e., twelve inches divided by pi inches, which equals 3.82.) Put another way, a circle with a circumference of 12 inches has a diameter of 12/pi inches; whether the circle has a circumference of 12 inches or 12 inches are added to the circumference of a larger circle does not matter. If you do the math for the earth example, you will find that the props must be 1.91 inches tall (1.91 is half of 3.82); the props are *half* of 3.82 because they are on both sides of the earth.

Now for the second surprise. Do the same problem with a sphere with a 9-inch circumference (which we will call a "baseball"), instead of with the earth (i.e., add a foot to the 9-inch taut string), so that the string is now 21 inches long. How tall must the props be?

You might think that this time the props will be much larger (because the foot of added string is much greater in relation to the baseball than the foot of added string in relation to the earth. You will be surprised to find that that the props must be the same, 1.91 . . . inches tall.

The reason for this is because this problem concerns only the relationship between the *addition* to the radius and the *addition* to the taut string. Nowhere does the size of the original sphere, be it earth or baseball, enter into the problem; that is, the lengths of the original diameter and the original circumference are irrelevant in each case. Of course, the 3.82 . . . inches added to the diameter and the foot that was added to the circumferences are much greater in proportion to a baseball than to the earth, but this has nothing to do with the question.

Answer #5: There are 292 ways to change a dollar—293 if you count a silver dollar as change.

When my nephew, Matt Mayers, was in high school, he gave the answer of 293. The teacher gave no credit for this answer, arguing that a silver dollar is not "change for a dollar." Even if this claim were not dubious in the extreme, the teacher's considering Matt's answer to be no better than no answer is strong evidence that this teacher is in the wrong field.

A computer method for finding the solution to this problem is given in *Creative Computing*, July–August 1977. A noncomputer method is given in G. Polya's *How to Solve It* (Princeton, NJ: Princeton University Press, 1957).

Answer #6: Switch. The car must be in one of only three places:

	Door 1	Door 2	Door 3
Possibility A	Car	Goat	Goat
Possibility B	Goat	Car	Goat
Possibility C	Goat	Goat	Car

Whichever door you select, Monte must open a (different) door with a goat behind it. (The rules Monte stated require this.) If the car is behind the door you first select, changing your selection to the other will turn a winning choice into a losing one. If the car is behind the other door, changing your selection will turn a losing choice into a winning one. Thus, you are twice as likely to win if you do change your selection as you are if you do not because there were two chances out of three that the car is behind one of the other two doors, and Monte has ruled out one of those two doors.

If you do not switch, your chances remain the same 1 in 3 they would be if Monte did not open a door with a goat behind it. They remain 1 in 3 because you ignore new information that improves the odds to 1 in 2.

Say you choose door 1 and do not switch. The car can be behind door 1, door 2, or door 3, so your chances of winning the car are the same 1 in 3 they would be if Monte did not open a door. You already knew that two doors had goats behind them, and Monte simply opened one of them. So the chances of winning do not change if you do not switch. Putting this another way, you are ignoring the information that reduces the odds from 1 in 3 to 1 in 2. (This would not be the case if Monte could open a door with the car behind it; but the problem stipulates that he is going to open one of the two doors with a goat behind it.)

Now, note that a new contestant, entering the game after Monte opens the door, chooses from only two doors, and therefore has a 1-in-2 chance of winning; knowing that the car is not behind the door that Monte opens, this new player will face odds of 1 in 2, not 1 in 3.) *You are in the same situation if you switch.*

Your chances of winning, then, *do* change if you switch. You become the equivalent of the "new player" just mentioned.

Say you chose door 1. The door Monte opened could be

A2, leaving A3 (goat) closed
A3, leaving A2 (goat) closed
B3, leaving B2 (car) closed
C2, leaving C3 (car) closed

Thus, when you switch you raise your chances of winning from 1 in 3 to 1 in 2 (i.e., 2 in 4).

Another Way of Looking at the Problem

The problem is sometimes explained in these simple terms: When you first chose a door, there was a one-third chance that door had the prize and a two-thirds chance that one of the other two doors had the prize. When Monte opened one of the other two doors, the remaining door—the one you can choose to switch to—has a two-thirds chance of being correct. While this is correct and simple, it usually fails to persuade because it is not intuitively clear.

Say that there are fifty doors, with a car behind one of them. This becomes more clear if we make two sets: (1) "Set 1," containing only one door, the one you randomly guessed to hide the car, and (2) "Set 2," containing the other forty-nine doors. Obviously, the odds are overwhelming that the car is behind one of the forty-nine doors in Set 2.

Now, Monte opens forty-eight of the forty-nine doors of Set 2; by the rules of the problem, none of these may have a car. This leaves two closed doors: the one you originally guessed and the remaining one of the forty-nine in Set 2.

The likelihood is still overwhelming (49/50) that the car is behind a door in Set 2. Because there is only one door left in Set 2, the likelihood is overwhelming that the car is behind this door and not behind the door you selected. Thus, you should change your selection. (You knew all along that there were at least forty-eight doors without cars in Set 2 and that Monte would have to open them; the 49/50 likelihood of having the car resided all along in the door in Set 2 that Monte would have to leave closed).

It is, of course, *possible* that you happened to choose the door with the car behind it and that Monte's forty-ninth door did not have a car behind it. But forty-nine out of fifty times (on average) this will not be the case.

The three-door problem is entirely analogous to this. However, because there are so few doors, it does not occur to us to think of the three-door problem in this way. It is the intuition that, whenever there are two doors, each must have a fifty-fifty chance that is responsible for so many people incorrectly assuming that the two of the three doors remaining in the three-door version each have a fifty-fifty chance of having the car.

The fifty-door analogy is helpful because it enables intellect to counter this incorrect intuition by making obvious the fact that Monte's opening the forty-eight doors gives us new information: it tells us that the door remaining in his set after he has opened forty-eight doors *still carries the original 49/50 probability* that the set of forty-nine had—much more obviously—before Monte opened the forty-eight doors.

In short, Monte *does* give us new information: one of the doors (the last remaining door of Monte's group) comes from a set with a higher probability of containing the winning door than does the door we selected at first. In the fifty-door version, the door we choose has a 1-in-50 chance of winning, while the remaining one of Monte's doors has a 49-in-50 chance. In the three-door version, the door we choose has a 1-in-3 chance, while the remaining of Monte's two doors has a 2-in-3 chance.

The three-door problem is entirely analogous to the fifty-door problem, but, because there are only three doors, the reality is much harder to see in the three-door version.

If you are still not persuaded, go to www.stat.sc.edu/west/javahtml/LetsMakeaDeal.html. There you can play the game many times per minute. Play for five minutes and the probability is overwhelming that you will win twice as often when you switch.

Answer #7: It cannot be done. To average sixty miles an hour for the two miles, you would have to complete the two miles in two minutes. You have already used two minutes in doing the first mile (1 mile at 30 mph = 2 min.). Unless you complete the second lap in no time at all, you cannot complete the two miles in the two minutes required for a sixty-mile-per-hour average for the entire race.

Answer #8: Obviously there must be at least three people at the party. But three are not sufficient, because, for example, one may know two, and three may know neither. If this is the case, then no three people know each other, and no three people do not know each other.

Four people are not sufficient because one may know only two, and three may know only four. If this is the case, then no three people know each other, and no three people do not know each other.

Five people are not sufficient because one may know only two and five; two may know only three and one; three may know only four and two; four may know only five and three; and five may know only one and four. If this is the case, then no three people know each other, and no three people do not know each other.

However, what if there are six people at dinner? Person one must either know at least three of the remaining five people or not know at least three of the five remaining people.

Now, here is an interesting fact: let's say we want to find out how many people must be present to guarantee that five people all know each other or five people do not all know each other. To guarantee that four people know each other or four people not know each other, there must be eighteen people at the party. To guarantee that four people know each other or five people do not know each other, there must be twenty-five people at the party. (The proof of this takes eleven years of desktop computer time.)

So how many to guarantee that six people know each other or do not know each other? No one knows; the answer is between forty-two and fifty-five persons.

All this is known as Ramsey Theory (for the seventy-year-old work of British mathematician Frank Ramsey; the leading current mathematician in the area—and many others—is Ronald Graham of AT&T Lab).

Answer #9: *Let's call the person who shook hands with no one "A," the person who shook hands with one person "B," and so on. (See the chart below.)*

A shook with	0 people
B	1
C	2
D	3
E	4
F	5
G	6
H	7
I	8

The central insight is this: If the person who shakes hands with eight people (I) is not the spouse of the person who shakes hands with no one (A), then there will be at least three people with whom the person who shakes hands with eight people (I) does not shake hands: himself/herself, *or* his/her spouse, and the person who shakes hands with no one (A). But this leaves only seven people, and makes shaking hands with eight people impossible. Therefore, the person who shakes hands with eight people (I) must be the person who is the spouse of the person who shakes hands with none (A).

Once you have had this insight, the rest follows easily, if not quickly. If the person who shakes hands with seven people (H) is not the spouse of the person who shakes hands with one person (B), then there are at least four people with whom the person who shakes hands with seven people (H) does not shake hands: himself/herself, his/her spouse, A (who shakes hands with no one), and B (the person who shakes only one person's hand; the one person must be the person who shakes the hands of all eight possible people (I). But this leaves only six people and makes shaking hands with seven people impossible. Therefore, the person who shakes hands with seven people (H) must be the spouse of the person who shakes hands with one (B). H cannot be the spouse of A because A is the spouse of I.

You continue until you discover that the person who shook hands with four people (E) must be the spouse of the anthropologist (who also shook the hands of four people).

	A	B	C	D	E	F	G	H	I	*
A	X	n	n	n	n	n	n	n	s	n
B	n	X	n	n	n	n	n	s	y	n
C	n	n	X	n	n	n	s	y	y	n
D	n	n	n	X	n	s	y	y	y	n
E	n	n	n	n	X	y	y	y	y	s
F	n	n	n	s	y	X	y	y	y	y
G	n	n	s	y	y	y	X	y	y	y
H	n	s	y	y	y	y	y	X	y	y
I	s	y	y	y	y	y	y	y	X	y
*	n	n	n	n	S	y	y	y	y	X

X = Self/n = Did not shake hands with/y = Shook hands * = Anthropologist with/s = Spouse

Answer #10: Intuition and common sense would have you reason that (A) the card on the table must be either the card that is white on both sides or the card that is white on one side and red on the other, and (B) there is one chance in two that the hidden side is white (if it is the card that is white on both sides) and one chance in two that the other side is red (if it is the card that is white on one side and red on the other). Thus, (C) as the odds are even that the other side is white, you should take the 3–2 odds.

You will not be shocked to learn that intuition and common sense are wrong. (After all, it would not be much of a brainteaser if they were not.)

Probability problems are notoriously counterintuitive because intuition fails to include some of the possibilities and compares events that are not equally probable. And probability calculation must always be based on equally probable events.

There are six possible situations in which the card on the table shows white:

Card Showing	1st Card in Hat	2nd Card in Hat
1. White shows White under	White on one side Red on other side	Red on one side Red on other side
2. White shows White under	Red on one side White on other side	Red on one side Red on other side
3. White shows White under	Red on one side Red on other side	White on one side Red on other side

4. White shows White under	Red on one side Red on other side	Red on one side White on other side
5. White shows Red under	White on one side White on other side	Red on one side Red on other side
6. White shows Red under	Red on one side Red on other	White on one side White on other side

In four of the six cases, the underside is white. Therefore, if you get 3–2 odds, you will lose because you will lose twice for every once you win. A third of the time you will win $1.50 for a $1.00 bet, and two-thirds of the time you will lose $1.00 on a $1.00 bet.

Answer #11: There is no coherent answer to this question because the question is incoherent. Say that the concierge discovered that the room should have cost twenty dollars and gave the bellhop ten singles to return to the men. Say the men took three dollars each and left the bellhop the remaining dollar as a tip.

It would not and should not occur to us to say: Thus, each man spent seven dollars, and the bellhop got one dollar, for a total of twenty-two dollars. Where did the other eight dollars go?

It is only because the numbers in the original problem are so close that we (incorrectly) expect these numbers to add up to the money originally paid by the men.

Answer #12: Ask either man, "If I were to ask you if your door is the door of death, would you say yes?"

If you are asking the truthful man and his door is the door of life, he will truthfully say that he would answer no.

If you are asking the truthful man and his door is the door of death, he will truthfully say that he would answer yes.

If you are asking the liar and his door is the door of life, he will lie and say that he would answer no. (That is, the liar would, in fact, lie and say, "Yes, this is the door of death." Because he would, in fact, say "yes," he answers your question about what he *would* say with a lying "no.")

If you are asking the liar and his door is the door of death, he will lie and say that he would answer yes. (That is, the liar would, in fact, lie and say, "No, this is the door of life." Because he would, in fact, say "no," he answers your question about what he *would* say with a lying "yes.")

Thus, a "yes" answer from either man means that his door is the door of death, and a "no" answer from either man means that his door is the door of life. So, if the man says "yes," choose the other man's door; if the man says "no," choose his door.

You never do find out whether the man who answered is the truth teller or a liar, but with the princess, the money, and the chance to meet Michael Jordan, who cares?

Answer #13: There are only two possibilities:

Miss Green wears blue, Miss Blue wears black, and Miss Black wears green.

Or

Miss Green wears black, Miss Blue wears green, and Miss Black wears blue.

Miss Blue cannot be wearing black because the woman wearing black says "So what" *to* Miss Blue. So, Miss Green wears black, Miss Blue wears green, and Miss Black wears blue.

Answer #14: Compare the offers over six-month periods:

	Offer A	Offer B
End of 6 months	50	
End of 12 months	55	100
Total 1 Year	105	100
End of 18 months	60	
End of 24 months	65	120
Total 2 years	230	220
End of 30 months	70	
End of 36 months	75	140
Total 3 years	375	360

Things continue in this way, so it is clear that A is the better choice.

Answer #15: No. Consider this: you win if you choose black-black-black and the first three spins are black-black-black. This will happen one-eighth of the time.* Any time after the first three spins, three blacks must be preceded by a red. But this means that the three spins before the last spin would have to have been red-black-black, the combination I chose to counter your black-black-black. Thus, I will win seven out of eight times (on average). The same strategy works whatever the combination you choose.

Some choices other than black-black-black improve your odds. But the best you (i.e., the person who goes first) can do is lower the odds to 2–1 against yourself.

*Because there are eight equally likely possibilities:

b-b-b
b-b-r
b-r-r
b-r-b
r-b-r
r-b-b
r-r-b
r-r-r

Answer #16: It is easy to conclude that, if a hundred chickens eat a hundred bushels in a hundred days, one chicken will eat one bushel in one day. This would imply that ten chickens will eat ten bushels in a day and a hundred chickens will eat a hundred bushels in ten days. Obviously something is wrong with this because we know that it takes a hundred chickens a hundred days to eat a hundred bushels.

The above reasoning is incorrect because if a hundred chickens eat a hundred bushels in a hundred days, then a hundred chickens eat one bushel in one day. Thus, ten chickens will eat one bushel in ten days. (And ten chickens will eat ten bushels in a hundred days. A hundred chickens will eat one bushel in one day, ten bushels in ten days, and a hundred bushels in a hundred days.)

Answer #17: This is a perfect example of the idea of symmetry. If you must place the cigar on its side, the person who goes second always wins (assuming that he plays correctly). The reason for this is that whenever the person who goes first places a cigar, the person who goes second places his cigar in precisely the "opposite" place on the table. Therefore, whenever the first player can place a cigar on the table, the second player can do so as well. (Note that it does not matter whether a cigar is permitted to be partly off the table; the same logic applies.)

However, if a player is permitted to stand a cigar on its flat end, then the first player always wins. The first player places his cigar, standing on the flat end, in the center of the table. This captures the one spot that has no symmetrical partner (it contains its own symmetry). By doing this, the first player effectively becomes, in the terms discussed above, the second player, and he always wins. (No, you cannot stand a cigar on its point. It is in the Bible. Look it up.)

Answer #18: The amount of red paint in barrel 1 equals the amount of blue paint in barrel 2. It seems that this should not be the case because you take undiluted red paint out of barrel 1 and put it in barrel 2 and then take the mixture out of barrel 2 and put it in barrel 1.

But look at it this way. The problem with that logic is that the 1 gallon from the red paint comes from 10 gallons. The diluted blue paint comes from 11 gallons, of which 10/11 is blue. There are 10 gallons in 10/11 of 11 gallons, the same amount as in the barrel of red paint at the start. Ten gallons of paint taken from 11 gallons of paint that is 10/11 blue leaves barrel 2 with the same amount of blue paint that barrel 1 has after its missing gallon is replaced with the 10 gallons of paint taken from barrel 2.

All numbers below refer to gallons

Barrel 1		Barrel 2
10 red	Start	10 blue
	1 gallon from barrel 1 to barrel 2	
9 red	Mix paint in barrel 2; Mixture is 10/11 blue and 1/11 red	10 blue + 1 red
		10/11 blue + 1/11 red
	1 gallon from barrel 2 to barrel 1	
9 red + 1 × 10/11 blue + 1 × 1/11 red		
9 + 1/11 red + 10/11 blue red		9 + 1/11 blue + 10/11

This problem easier to understand if we consider an analogous problem that appeared in Marilyn vos Savant's column, Ask Marilyn, in *PARADE* magazine. Say there is a boys' college right next to a girls' college. Each has a dorm that holds 1,000 people. One night, 500 girls raid the boys' dorm. At this point, there are 1,500 students stuffed into the boys' dorm. The fire marshal arrives and orders (any) 500 of the 1,500 to go to the girls' dorm. Some girls stay and others leave. Which dorm now has the greater concentration of its original sex?

In a situation comparable to that in the paint problem, each has the same number of the "other sex" students and the same number of "original" students. Because each must end up with 1,000 students, for every girl who stays in the boys' dorm (i.e., every girl who is not one of the 500 students who go to the girls' dorm), there must be a boy who goes to the girls' dorm (i.e., every boy who *is* one of the 500). Thus, each dorm will end up with the same number of "other sex" students.

Marilyn vos Savant also provides another version of this. You are blindfolded and given a deck of cards with (any) ten cards facing up and the other forty-two facing down. You must divide the deck into two stacks, each containing the same number of cards facing up. Can you do this?

Yes. Simply deal yourself ten cards and turn all of them over. If, say, none of the first ten cards of the original deck had been facing up, then the remaining forty-two card stack will have ten cards facing up, as will your ten-card stack (because the ten cards you dealt yourself would have no up-facing cards before you turn them over and, therefore, ten up-facing cards when you turn them over). If one of the first ten cards of the original deck had been facing up, there will now be nine cards facing up in the forty-two card stack and nine facing up in the ten-card stack.

Answer #19: There is no space between the poles; they are touching. If the lowest point of the rope is twenty feet off the ground, the rope must go down thirty feet on one pole and up thirty feet on the other. That takes care of the rope's sixty feet, so the poles must be next to each other. (You are also correct if you answered that this situation cannot exist. You can reasonably argue that, even if the poles are touching, a slight bit of rope must be taken by the distance between the two attachment points, in which case there would not be quite enough rope to go down and up sixty feet—which it must if it is to reach a point only twenty feet from the ground.)

Answer #20: Forget all those equations, formulae, and diagrams you have just committed to paper. Instead, picture *two* monks, one starting up at the same time that other starts down. Obviously, they must meet at some point.

Answer #21: Light one piece of string at *both* ends. Light the other piece at one end. Whatever the rate of burning of the first string, the two flames will meet after half an hour. (You do not know where they will meet, but it does not matter.)

When the two flames of the first string meet, light the second end of the second piece of string. In fifteen minutes, the two flames of the second piece will meet. The burning rates of the two flames of the second piece of string do not matter; the two flames burning on the second string will meet in fifteen minutes.

Thirty minutes plus fifteen minutes is forty-five minutes.

Answer #22:

Tom	Harry		You
R	W	Tom raises his hand. As Harry's hat is white, you must be the red Tom sees.	R
W	R	Harry raises his hand. As Tom's hat is white, you must be the red Tom sees.	R
W	W	Tom and Harry raise their hands, so you must be red.	R
R	R	Tom and Harry each knows the other has a red hat, and they both raise their hands. But if you had a white hat, both Tom and Harry would know they had the red hat that the other saw and each would raise his hand, so you could not have a white hat.	W
R	W	Tom does not raise his hand, so you, like Harry, must be white.	W
W	R	Harry does not raise his hand, so you, like Tom, must be white	W
W	W	Tom and Harry do not raise their hands, so you must be white.	W
R	R	Tom and Harry raise their hands but do not say the color of their own hats. If you had a white hat, each of them would have known that he had the red hat the other saw and would have figured out that he had a red hat. So you must have a red hat. (Of course, Tom and Harry each had all the evidence you did, but you are quicker.)	R

Note that whomever of the three is quickest can tell the color of his hat no matter what color of any of the hats.

Answer #23: Obviously, you cannot simply ask, "Which path leads to heaven?" You do not know whether the answer is coming from the truth teller or the liar.

But how about trying this. Let's assume the path to heaven, the one you would like to take, is the west path. You ask either man, "If I ask the *other* man which path is the path to heaven, what will he say?"

If the man you happen to ask is the truth teller, meaning the other man is the liar, he will truthfully answer, "East." (That is, he will truthfully tell you that the liar will lie and tell you "east.")

If the man you happen to ask is the liar, he will lie and say, "East." (That is, he will lie and tell you that the truth teller would tell you "east.")

You still do not know which man is the truth teller and which is the liar, but you do not care. You do know what you wanted to know: the west path is the path to heaven. (Of course, this works whichever one is the path to heaven.)

Answer #24: You cannot. Think in terms of two people, rather than just you. The one who leaves later must catch up to the one who leaves earlier. At that point, they meet the same traffic and arrive at work at the same time. In other words, the best you can do by leaving later is to arrive at the same time (which, of course, is a good idea when you can in fact catch up to the earlier guy).

Answer #25: The downtown train comes at one minute after the hour and every six minutes after that. The downtown train comes at two minutes after the hour and every six minutes after that. So, unless you get to the subway within the minute after an uptown train has come, you will always get the uptown train.

Answers to Nonbrainteasers in the Text

Note: I used to include among my favorite brainteasers the following:

> You have a box six feet by six feet by six feet. You want to fill it with balls in such a way that the least empty space is left. All the balls must be the same (finite) size and whole (i.e., no parts of balls). For example, you may fill the box with one ball with a six-foot diameter or four balls with three-foot diameters, and so on.
>
> What size ball(s) would you choose?

It used to be thought that the correct answer is that it makes no difference; there is always the same amount of *total* empty space left. This answer was surprising to most people and fun—in other words, a fine brainteaser. Alas, some mathematician pointed out this is true only if one of a number of possible methods of packing is employed. Including this constraint in the question makes for good mathematics—and a brainteaser that is incomprehensible to the nonmathematicians for whom brainteasers are meant.

Answer for the fireman-brakeman-engineer problem: The brakeman, who lives halfway between Detroit and Chicago, also lives nearest Mr. X. Mr. X earns exactly three times as much as the brakeman. Mr. X cannot be Mr. Robinson because Mr. Robinson lives in Detroit (and therefore could not be the brakeman's nearest neighbor). Mr. X cannot be Mr. Jones because Mr. Jones earns $20,000 a year, which could not be *exactly* three times the brakeman's salary.

Because Mr. X cannot be Mr. Robinson or Mr. Jones, he must be Mr. Smith.

The passenger whose name is the same as the brakeman's lives in Chicago. Therefore, he cannot be Mr. Robinson, who lives in Detroit. He cannot be Mr. Smith because Mr. Smith lives nearest the brakeman, who lives halfway between Detroit and Chicago.

Because the passenger whose name is the same as the brakeman's cannot be Mr. Robinson or Mr. Smith, he must be Mr. Jones. Because his name is the same as the brakeman's, the brakeman must be Jones.

Smith beats the fireman at billiards, so the fireman cannot be Smith. The fireman cannot be Jones because Jones is the brakeman. So Smith must be the engineer.

If the brakeman is Jones and the engineer is Smith, the fireman must be Robinson.

Notes

1. “Knowing” and “not knowing” are mutual (i.e., if one person knows a second, then the second knows the first, and if one person does not know a second, then the second does not know the first).
2. “Three people all know each other” means that *each* of the three must know the other *two.* It is not sufficient for A to know B and B to know C. C must also know A. Likewise, “three people all do not know each other” means that *none* of the three know *any* other of the three.
3. *Games* magazine (January–February, 1978).
4. This assumes a flat surface. When we talk about a surface that is not flat, it is a whole different ball game. Here we are dealing with a flat surface because the circumference of the earth or baseball is a circle on a plane. If we were to draw a circle *on* the earth or the baseball, then plane geometry would not apply.

Index